# Telling Stories with Maps

# TELLING STORIES WITH MAPS

Lessons from a lifetime of creating place-based narratives

## Allen Carroll

**Esri Press**
Redlands, California

Esri Press, 380 New York Street, Redlands, California 92373-8100
Copyright © 2025 Esri
All rights reserved.
Printed in the United States of America.

29 28 27 26 25        3 4 5 6 7 8 9 10

ISBN: 9781589487970
Library of Congress Control Number: 2024951472

# Contents

Foreword                                                          vii

Introduction                                                       ix

**Chapter 1:** Why Stories Matter                                    1

**Chapter 2:** Why Maps Matter                                      33

**Chapter 3:** Maps and Minds                                       51

**Chapter 4:** From Analog to Digital                               67

**Chapter 5:** The Journey to Storytelling                          91

**Chapter 6:** Maps in Dramatic Roles                              123

**Chapter 7:** Nine Steps to Great Storytelling                    145

**Chapter 8:** Plan, Produce, Polish, Publish                      167

Epilogue                                                           206

Acknowledgments                                                    213

References                                                         215

Credits                                                            219

Visit the ArcGIS StoryMaps stories marked in this book with a ⊖ at

**go.esri.com/telling-stories-site**.

# Dedication

To Jack and Laura Dangermond, whose vision is changing the world

To Clint Brown, whose leadership is revolutionizing GIS

# Foreword

Maps. They have been—and still are—a big part of my life.

Today, I live in Melbourne, Australia, for part of each year. And London, England, for another part. And all sorts of other places around the world in between.

If my life is still a movable feast (and I like it that way), it's probably because that's how it started. I was born in England, but when I was still only a year old, my family moved to Pakistan, and for the rest of my childhood we just kept moving. I can blame my father for that. I'm a prime example of the baby boomer generation, born right after World War II. My father, an RAF pilot during the war, transferred to British Overseas Airways Corporation (BOAC) just before I arrived. For 30-plus years, BOAC was the interval between Imperial Airways and today's British Airways. His airline occupation kept him shifting around, and, as children do, I tagged along. Every two years, it seemed to be another school: After Karachi in Pakistan, it was Nassau in the Bahamas, then a very brief break back in England, followed by Detroit, Michigan (back when Motor City was spelt with big capital letters and tail fins soared toward the sky), and then another American spell in Baltimore, Maryland.

Just to prove my American Midwest, East Coast, and West Coast credentials, much later on, by this time with my own children, I also lived in the San Francisco Bay area—in Berkeley, to be precise.

That nomadic childhood certainly gave me an early enthusiasm for maps. My children regularly proclaimed that it was absurd that any kid would have asked Santa for a globe and later for a filing cabinet to keep all that mapping information properly organised. Then, after ending up back in England and going to university, my just-married wife and I set out on a little drive along the "hippie trail," a road trip that took us all the way from Europe to Afghanistan and then on through India and Southeast Asia to Australia, where we started Lonely Planet to publish guidebooks about our preferred sort of travel.

I could say my career proceeded to follow a remarkably similar track to Allen Carroll's, as he documents in *Telling Stories with Maps*. Back when maps were lines on paper and print rather than digital, giving people important information was what our respective lives were all about. I don't think National Geographic maps were telling their users to "turn right out of the train station and you'll quickly find a good, cheap hotel. Turn left and you'll get mugged," but otherwise the information our maps provided was equally vital.

Like Allen, I was fortunate to see mapping go through huge and often challenging revolutions. In my early guidebook mapping days, finding a mapping baseline as a starting point was often a major trial. For our first India guidebook, going back to something produced by the British colonial government in the era of Queen Victoria was often a better bet than relying on anything the Indian government was able to provide. Today, of course, that has all changed—we can instantly pinpoint anywhere in the world with a satellite view that can even identify an automobile in the street. For a spell, if I checked out my mother's home on Google Earth, it was clear that the satellite had passed over on a Saturday because the image showed my yellow sports car in her driveway. If I was in the country, I always tried to drop by to have lunch with her on Saturdays.

So my guidebook mapping story shifted dramatically over the years, as Allen's did as he moved from National Geographic to Esri. But although today I'm a map user far more than a map creator, Allen's story is definitely ongoing. ArcGIS® StoryMaps℠ stories today tell a far bigger and better tale than anything either of us produced in the print-and-paper mapping era. I love the concept that a map today can be choreographed, it can sing and dance, and for me that's what mapping is still all about. Out of that train station, left or right, and I'm on my way, racing down the road to wherever the map might take me.

Tony Wheeler
Writer, cofounder of Lonely Planet, and world traveler

# Introduction

## A spatial kid

That's me, in my pj's, in the middle of the picture. I was four or five years old. I'm with my grandmother and my brother. In the background is a world map I remember staring at from my bunk bed. I loved its patterns and colors. I imagined nudging South America and Africa together in my mind. I wasn't aware of it, but Africa in the mid-20th century was mostly partitioned into European colonies; a historical geographer could probably age me by studying its details.

That large-format, world political wall map made a deep impression on me. It's likely that it spawned some neural pathways in my young mind that would shape my later life.

I hope you'll bear with me while I introduce myself and briefly describe my childhood and youth. I think it might add some context to this book, and help explain my peculiar passion for maps, geography, and storytelling. In a larger sense, I think my experience might serve as a reminder that we need to recognize and nurture the instinctive passion that so many children have for maps and geographic exploration. In a sense, that's what this book is about: inspiring people of all ages to explore modern mapping and storytelling technologies that make us more aware of the world and more conscientious as world citizens.

A few years after the pajama snapshot was taken, I assumed the role of family navigator. My parents, in classic 1950s and 60s fashion, piled us into their Pontiac station wagon, departing from Indianapolis every summer on long driving vacations punctuated by overnights in Holiday Inns and Howard Johnson's. I would plot our routes ahead of time. The interstate highway system was a work in progress back then; many of our drives alternated between freeway sprints and slow slogs through Midwestern county seats. I'd conscientiously track our progress, the *Rand McNally Road Atlas* open on my lap.

As a teenager, I took long bike rides around the north side of Indianapolis, cycling down as many blocks as possible and marking my accomplishments on a Shell Oil street map exactly like the one at right. From the age

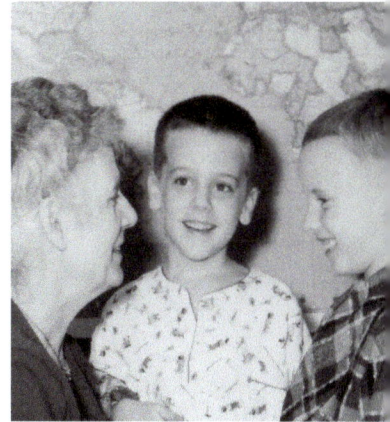

My grandmother, me, my brother, and the world.

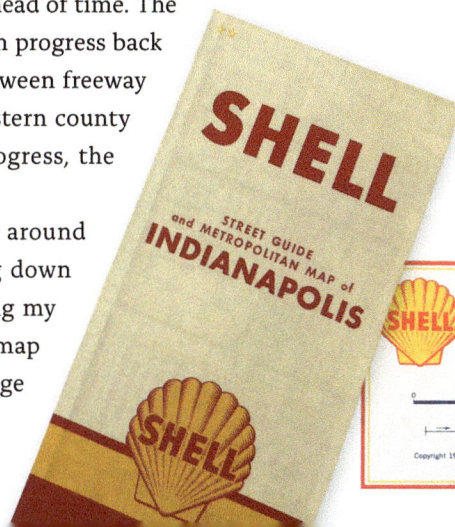

SHELL
STREET GUIDE and METROPOLITAN MAP of INDIANAPOLIS
SHELL

A Shell 1960 street map of Indianapolis.

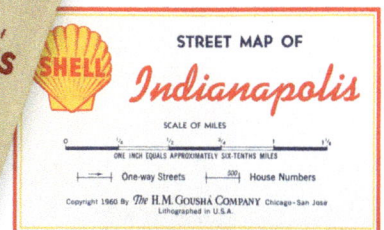

STREET MAP OF
*Indianapolis*
SCALE OF MILES
ONE INCH EQUALS APPROXIMATELY SIX-TENTHS MILES
One-way Streets    House Numbers
Copyright 1960 By *The* H.M. GOUSHÁ COMPANY Chicago-San Jose
Lithographed in U.S.A.

of about eight and into my high school years, I was sure I was destined to be an architect. I loved buildings, and I loved to draw and make floor plans—which are analogous to maps, of course. I'm sure it never occurred to me that I might make a living making maps.

My interests began to evolve in my teen years. My high school biology teacher required us to spend time outdoors identifying birds. Thanks to her, birds captured my imagination; they opened my eyes to the world. I had been almost entirely unaware that these beautiful creatures were all around me. I had found a new passion, for birds specifically and the natural world more broadly. That discovery happened to have come at a time when the environmental movement was gaining momentum, inspired in part by the 1962 publication of *Silent Spring* by Rachel Carson. I left the Midwest to begin my undergraduate studies at Connecticut College and before long settled on human ecology for my major. The program was among the first environmental studies programs in the nation. I also began writing articles for the weekly student newspaper. During my junior year I became its editor, which for me was the equivalent of having a second major in journalism. I learned an enormous amount about reporting, writing, editing, design, and production—all with no notion that a digital revolution, a couple of decades in the future, would change everything.

I also spent many hours birdwatching in the Connecticut College Arboretum, 750 acres of woods and wetlands surrounding the campus. I hand drew the map of the arboretum on the facing page a year or two after I graduated as a sort of thank-you.

A few months after graduating—this was 1973—I took a job as an analyst for the Connecticut Department of Environmental Protection (DEP). Although I learned relatively quickly that I wasn't cut out to be a state bureaucrat, I happily became involved in producing publications for the department.

At this point, I was living in a tiny cottage on the land of my college adviser and mentor, Richard H. Goodwin. He was one of the founders of The Nature Conservancy, which has since grown to rank among the world's largest conservation organizations. His property forms the heart of what is now the Conservancy's Burnham Brook Preserve, 1,100 acres of second-growth woodland, pastures, granite outcrops, and watercourses. I lived in this pastoral environment for five years, during which I conducted surveys of breeding birds, getting to know the local community of woodcocks, warblers, thrushes, and whippoorwills.

Meanwhile, I left DEP and was recruited to become the editor of my alma

Burnham Brook Preserve, East Haddam, Connecticut.

mater's alumni magazine. It was a shoestring operation; I did much of the writing and all the designing, editing, and illustration.

At this point, I was contemplating my future career path. I could become a writer and editor, or I could become a designer and illustrator. I gradually concluded that writing and editing were hard work, and that designing and illustrating were fun—and perhaps a little more marketable. I assigned myself illustration tasks for the alumni magazine, which helped me assemble a portfolio.

Map of the Connecticut College Arboretum showing, with hand-stippled patterns, the woods and wetlands surrounding the campus.

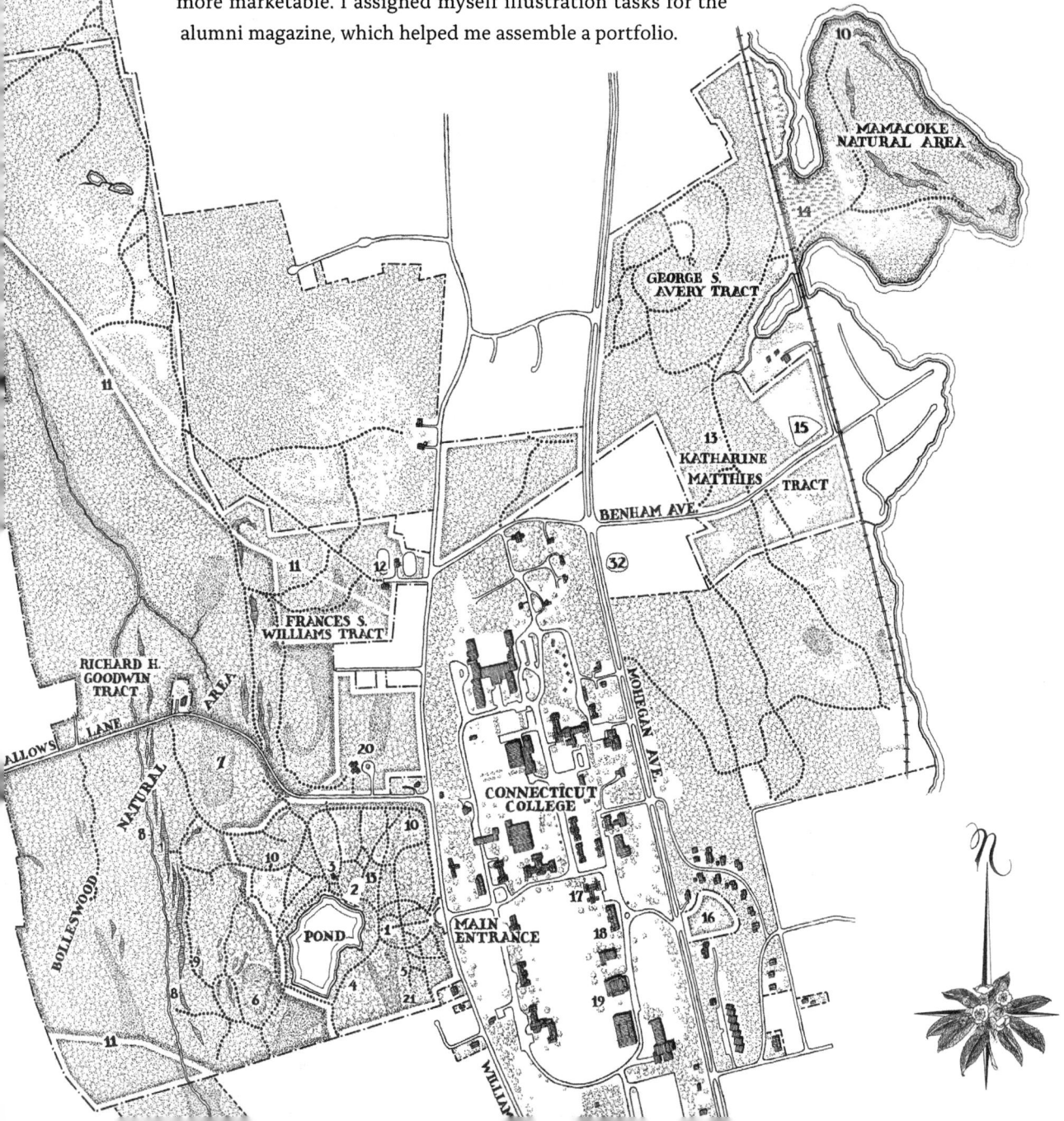

I loved living in the countryside, but I was lonely. I should perhaps have moved to New York City, with its vibrant publishing industry, but the notion of the big city scared me. My brother had been living in Washington, DC, and it seemed like friendlier and more familiar territory. So in May 1978, I packed up my portfolio and moved south. Despite my fear of making cold calls to art directors, within a few months I had accumulated a growing list of clients, including Johns Hopkins University in nearby Baltimore, *The New Republic*, the American Film Institute, the Council on Environmental Quality, the long-defunct *Washington Star*, and *The Washington Post*. I freelanced for about five years and thoroughly enjoyed it. The only position I could imagine enjoying more was joining the staff of National Geographic. I had made a couple of attempts to show my portfolio there but to no avail.

Meanwhile, I had befriended Ed Schneider, the art director of the now-defunct *Washington Post Sunday Magazine*. He invited me to come up with ideas for visual articles to occupy the center spread of the publication.

I pitched the concept of showing what's beneath Dupont Circle near downtown Washington: its utility infrastructure, transit tunnels, layers of soil and bedrock, and geologic setting. He approved, and I began to research the piece. I was having trouble unearthing information about precise locations until the chief engineer of the DC Metro System took me under his wing, sharing detailed plans (something that's unthinkable today—it was 1981, a more innocent age), and even walking me through the abandoned streetcar tunnels beneath the circle's busy traffic. After multiple sketches, I came up with a six-step infographic that I painstakingly executed using airbrush and colored pencils. It started with a small block showing the layers of pavement beneath the street; then a large-scale cutaway of the circle, exposing tunnels, Metro station, and utility lines; then a split block showing the Piedmont and Tidewater geologic provinces, on whose boundary Washington sits. Then came a cutaway of Earth, showing crust, mantle, outer core, and inner core. And, finally, a map (*facing page, top*) showing the location on Earth's surface opposite Dupont Circle—in a remote region of the eastern Indian Ocean. The Post received two letters to the editor claiming that I got the antipodal location wrong—but the letters insisted on two different locations!

From a business point of view, I lavished far too much time and attention on the Dupont Circle project, given the Post's modest payment policies. But I didn't mind; it was a labor of love.

The illustration strikes me now as embodying, in analog form, several of

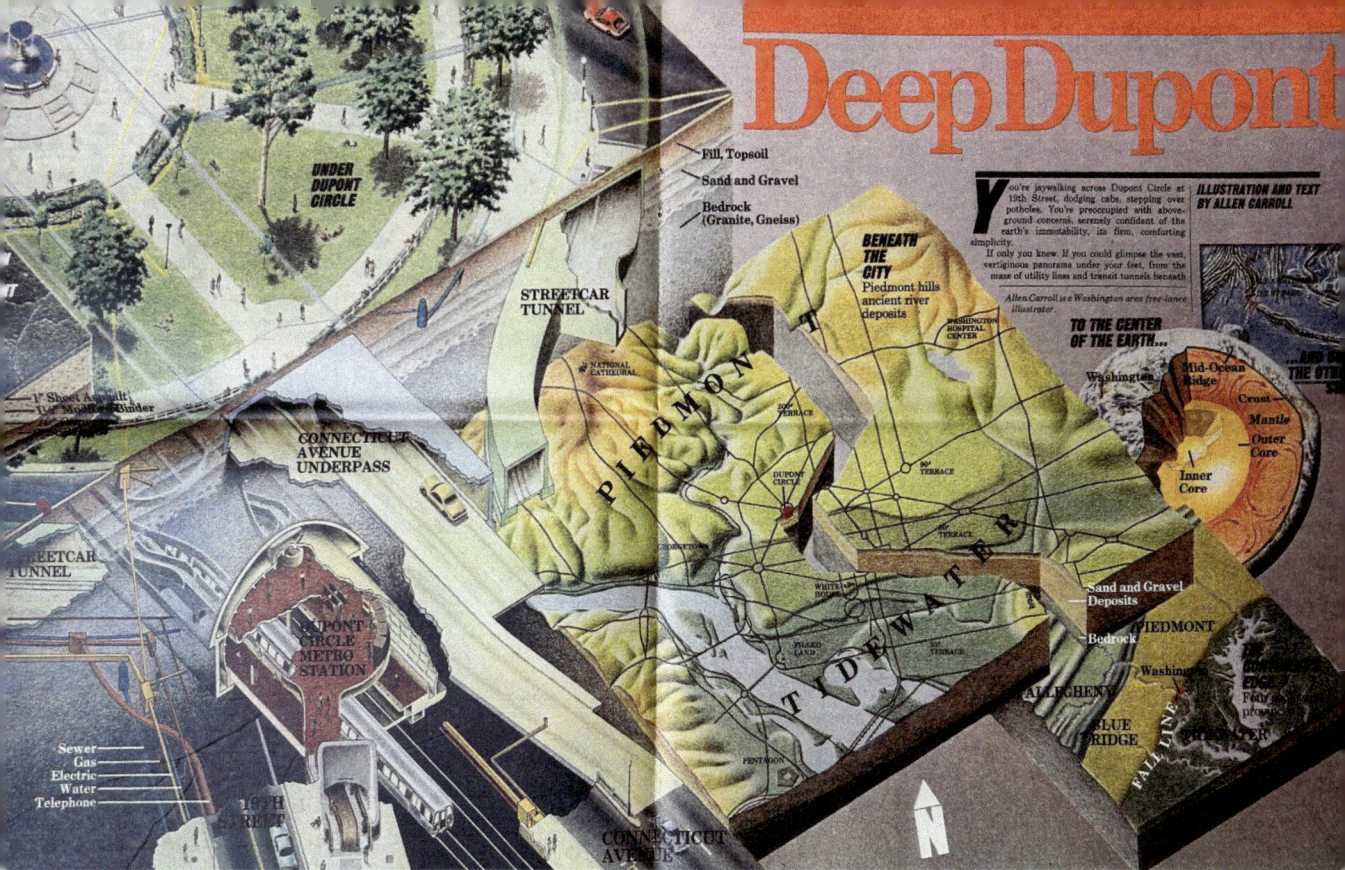

Deep Dupont

You're jaywalking across Dupont Circle at 19th Street, dodging cabs, stepping over potholes. You're preoccupied with above-ground concerns, serenely confident of the earth's immutability, its firm, comforting simplicity.

If only you knew. If you could glimpse the vast, vertiginous panorama under your feet, from the maze of utility lines and transit tunnels beneath

*Allen Carroll is a Washington area free-lance illustrator.*

ILLUSTRATION AND TEXT BY ALLEN CARROLL

the elements that make for a successful location-based multimedia story: An intriguing topic, a clear progression of elements, a variety of scales, a logical narrative sequence (although Ed had me do a clarifying key drawing, *right*), and a fresh look at a familiar landscape. It's not hard to imagine how this topic might be presented as an online multimedia narrative, with a mix of photographs, maps, 3D models, and animations.

This story has a postlude: I showed my illustration portfolio, including "Deep Dupont," to Jane D'Alelio, a Washington-based graphic designer. She thought my piece might be of interest to her husband, who happened to be Howard Paine, art director of *National Geographic Magazine*. I showed him my work, and he was excited about the Dupont piece. A few freelance assignments later, I was invited to work under contract for the magazine (this was spring 1983), and by year's end, I was a permanent, full-time assistant art director at the National Geographic Society, with Howard as mentor. I guess the time was well spent after all.

I went on to work at National Geographic for 27 exciting years before joining Esri in late 2010. You'll find brief autobiographical passages about this more recent history in chapters 4 and 5.

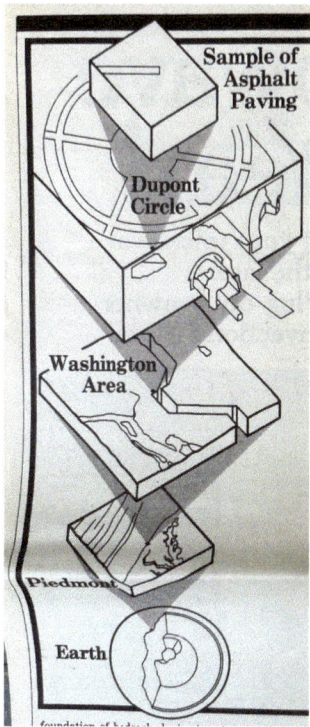

# About this book

The details of my less-than-scintillating history matter less on their own than they do as a device for contemplating how visual and place-based storytelling has evolved from the final decades of the analog age to our online, interconnected, digital present. Over my career, I've had plenty of time to think about telling stories with maps. I've written about it on a number of occasions, and I've created scores of stories myself. Some of this book's chapters are adaptations of articles, blog posts, and place-based stories I've produced over the years. Thus the pages of this book will take you on a winding journey, from speculation on how our brains process spatial information to practical tips on creating stories.

Chapter 1 ("Why Stories Matter") considers how storytelling is fundamental to our humanity and attempts to put the tremendous variety of place-based topics and approaches embraced by multimedia storytellers into 15 categories.

Chapter 2 ("Why Maps Matter") contemplates how maps help us tackle the *who, what, when, where, why,* and *how*—questions at the heart of understanding a topic or issue. It also compares and contrasts how words, images, and maps help facilitate answering the "five W's and one H."

Chapter 3 ("Maps and Minds") considers the deep connection between maps and memory. It explores the structures and functions within our brains that enable us to navigate the world and how exposure to maps and geography can change how we think.

Chapter 4 ("From Analog to Digital") explains how cartography and location-based storytelling have changed during the advent of the digital age and how my colleagues and I developed storytelling approaches that were suited to the new medium of the web while continuing to honor the centuries-old conventions of cartography.

Chapter 5 ("The Journey to Storytelling") recounts how my collaborators and I employed and refined storytelling techniques to create printed wall maps and atlas plates, and more recently to develop multimedia stories presented within the confines of small screens but liberated by a new ability to employ motion and interactivity.

Chapter 6 ("Maps in Dramatic Roles") categorizes the various functions that maps play in narrative contexts, from bit parts—simple locator maps—to stars of the show—maps that pan, zoom, and morph to interpret complex spatial data.

**Chapter 7 ("Nine Steps to Great Storytelling")** provides advice on how to approach and create effective stories, based on my team's long experience in producing place-based narratives.

**Chapter 8 ("Plan, Produce, Polish, Publish")** takes readers through the features and functions of ArcGIS StoryMaps, Esri's storytelling platform, offering advice and opinions along the way.

Between chapters are visual summaries of a few exemplary stories, some developed by myself and my team, others created by our growing community of storytellers.

The epilogue is my opportunity to speculate on the future of place-based stories and offer some thoughts about their value and meaning.

A challenge in writing this book lay in how to depict a moving target. Storytelling techniques will continue to evolve—we're only just past dawn in the digital age—and ArcGIS StoryMaps will continue to add new features. I urge you to explore the book's companion website (go.esri.com /telling-stories-site), which includes all the stories marked with a link symbol (⇔) mentioned in these pages. More importantly, please explore the many tutorials, instructional stories, blog posts, galleries, and videos on the ArcGIS StoryMaps website (https://storymaps.arcgis.com).

Finally, if you haven't already, please try using our software to tell your own stories. Please rediscover and indulge your own sense of wonder at the world that I suspect you, too, experienced as a child. I have no doubt that you have important tales to tell. Please add your threads to the tapestry— to the story of the Earth.

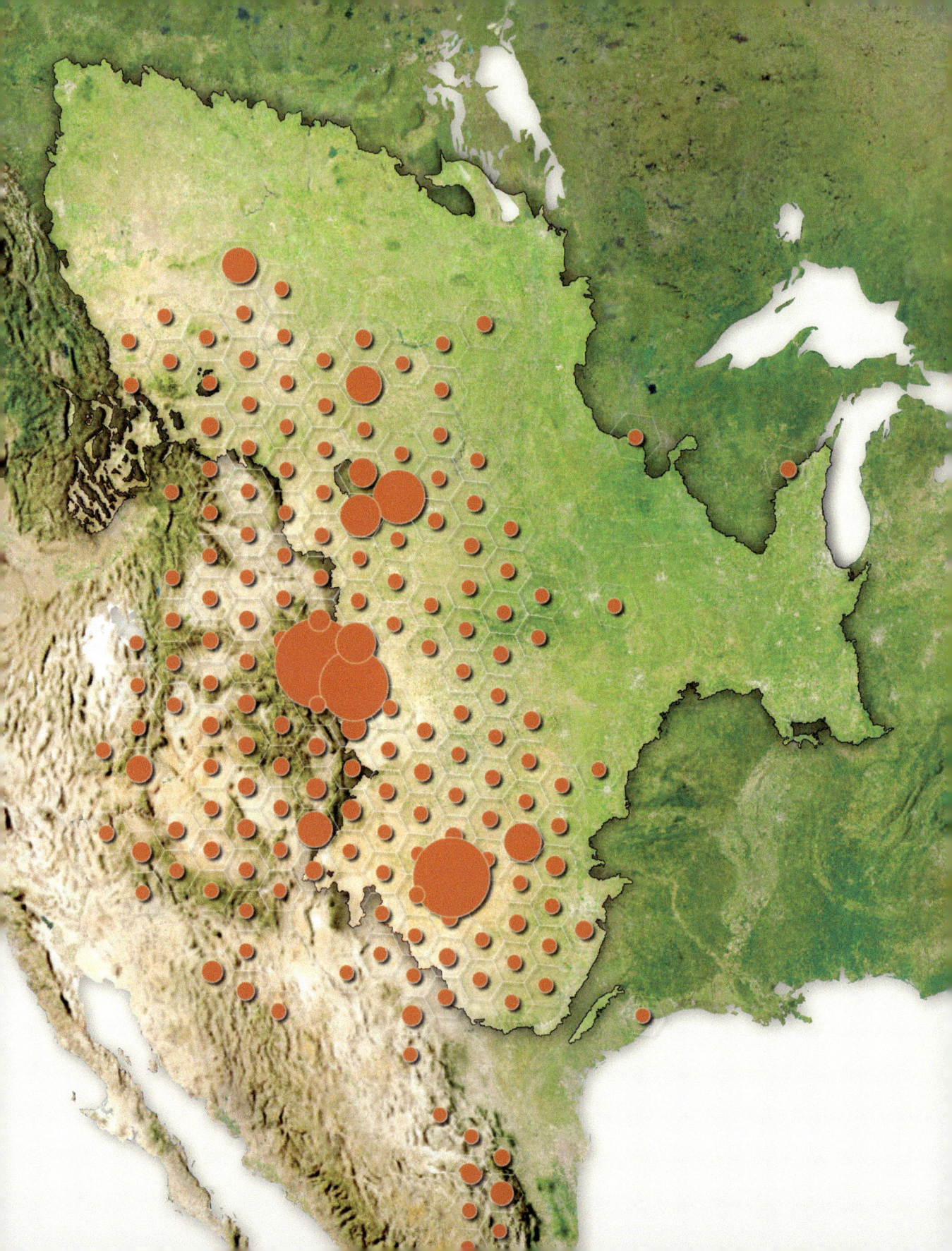

# 1

## Why Stories Matter

Prairie dogs play a pivotal role in maintaining the delicate balance of the grassland biome. These ecosystem engineers provide many crucial services to their environment that far exceed their small size. Thousands of iNaturalist observations, visualized here, indicate the prairie dogs' broad distribution.

## The allure of stories

Humans have told stories for thousands of years. Stories are the essence of how we communicate and how we form and reinforce our social ties. Stories help us make sense of the world and assign meaning to the often chaotic and confusing experiences of our lives. Until quite recently in human history, oral storytelling was the only way to pass down knowledge, beliefs, and traditions from generation to generation. So it should come as no surprise that storytelling is wound deeply into our cultural—and perhaps even our actual—DNA.

*The Storyteller*, 1773, Giovanni Domenico Tiepolo.

To some extent, we all have an innate talent for distilling the endless complexity of our experience into narratives that edit out the irrelevant and add our own reactions and conclusions. Every moment of our waking lives, we experience an onslaught of endless detail—an unceasing flood of visual, aural, and tactile stimuli. At a very young age, we learn to parse these stimuli, focusing on the small subset of inputs that enable us to move through space, make decisions, and interact with one another.

This unconscious editing process continues after the fact. Our minds process our experiences and store a pared-down record of prior events. As an experience recedes further into the past, the editing process continues until whole days and weeks of our lives are deleted from our memories. But exciting, terrifying, and memorable events remain vivid. When we recall and articulate these memories, in conversations with friends or in writing, we further organize and distill. We remove extraneous detail.

But telling stories is about much more than condensing and paraphrasing. Stories convey information, but they also convey emotion. Our recollections pass through the filter of our psyches; memories lose detail even as they gain emotional nuance. We recall events by way of our mind's eye, but that metaphorical eye's vision is subjective. We blend our emotional responses, our subjective interpretation of past events, into our accounts. And, yes, we sometimes exaggerate and embellish. We tell *stories*.

Stories are also about empathy—about relating to the experience of the storyteller by imagining being in that person's position and sharing his or her emotional journey.

The array of stories we tell and the many names we've given them show the importance of stories to our lives. Here are a few: account, allegory, anecdote, chronicle, diary, fable, lesson, memoir, myth, narrative, novel, parable, recollection, reminiscence, sermon, tale, yarn.

And then there are the *settings* within which we tell stories. Out of a long history of oral storytelling has evolved an array of storytelling formats and media: oratory, plays, operas, radio broadcasts, film, television, and, more recently, social media, podcasts, and web-based multimedia.

Throughout this evolution, another storytelling medium has evolved and persisted—one that, before the digital revolution, communicated through a visual language that is quieter and more subtle than most and that uses a largely visual vocabulary to tell stories about the world. That medium is maps. Maps in the predigital age were largely static and told their stories using subtle design and typographic techniques, emphasizing features by creating visual hierarchies. They also told stories by their choices of what information to include—or not include.

The quotation at right used to adorn the entrance to the National Geographic Society's map division. I saw it hundreds of times before it finally began to dawn on me: *Whose* dreams were those maps depicting? Maps of his era—and ours, for that matter—reflect conquest and colonial occupation that largely ignored or eradicated Indigenous cultures and boundaries. How many *unrealized* dreams do our maps ignore?

Dreams aside, I often think of printed or predigital-era maps as *nouns* because of their static, unchanging nature. As Esri® founder Jack Dangermond points out in his book, *The Power of Where*, "Today, *map* is increasingly

> A map is the greatest of all epic poems. Its lines and colors show the realization of great dreams.
>
> **Gilbert H. Grosvenor**
> Founding editor, *National Geographic Magazine*

an active verb." Maps pan, zoom, dance, and weave, revealing and parsing our dramatically changing world. They've also been freed to join other media—photos, video, audio, text, and so on—to perform as equals in multimedia narratives. I've been fortunate, in my roles at National Geographic and Esri, to work with many talented people trying to figure out how maps can help tell stories in this exciting new realm.

Part of that process has involved thinking about the types and structures of stories themselves and how maps can perform various functions within these structures.

## The shapes of stories

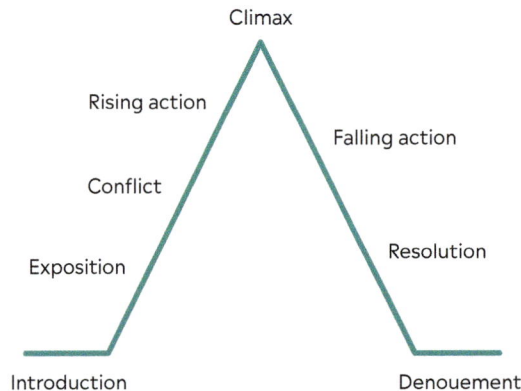

The authors of books and articles about storytelling have made many attempts over the years to diagram story structures as squiggles, peaks, pyramids, mountain ranges, wheels, grids, and so on. Some of them resemble the sketches at left.

One of my favorite discourses on the shapes of stories is a lecture given many years ago (and easily found on the internet) by the author Kurt Vonnegut, who stands in front of a blackboard and, in classic Vonnegut style, reveals insights through humor. He diagrams three stories, including Cinderella's circuitous journey from domestic misery to happily ever after. The curves he draws follow very different trajectories.

The lesson to me is that stories can take many forms, and no simple principle can be applied universally to the structure of stories and storytelling, with the possible exception that all stories have a beginning, a middle, and an end.

Many of the diagrams share the following characteristics.

The shapes of stories.

But do map-driven multimedia narratives inevitably include rising action, a climax, and falling action? No. Many of the books about storytelling, and many of the attempts to diagram story structure, are created with novels, screenplays, or oral storytelling in mind. Maps and the web are very different mediums from stage, movie screen, or printed page. Some of the same principles apply across these various mediums, but others don't.

Another factor is how one defines *story*. Is a walking tour of a historic district a story? Some would say no. But a tour has a beginning, a middle, and an end. And presenting the tour as a multimedia narrative means that it's likely to have some sort of title, an introduction, and a conclusion that might include links to additional information. To me, that's a story, regardless of the presence or absence of rising action, climax, and falling action.

Let's agree, then, that location-based multimedia narratives are, in fact, stories, and that those stories take different forms. All of them have a beginning, a middle, and an end:

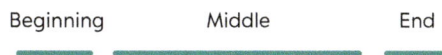

Beginning     Middle     End

In most cases the beginning includes a title and introduction, the middle is some form of exposition, and the end is a conclusion and/or call to action, if only a link to more information. The exposition, or story arc, in between might describe a research project or a spatial analysis:

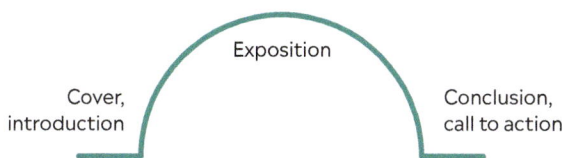

Exposition

Cover,
introduction

Conclusion,
call to action

Or it might be a guided tour taking readers on a virtual journey from point to point through a neighborhood or across a continent:

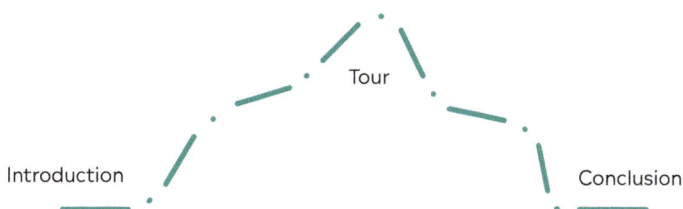

Tour

Introduction

Conclusion

Another story could describe a problem or issue, then demonstrate what an organization is doing to resolve or alleviate the issue. This scribble has a familiar look:

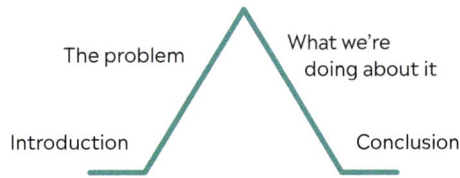

Other stories present an issue or problem in microcosm or as a case study, then provide an overview to show that the microcosm is symptomatic of a regional or global phenomenon:

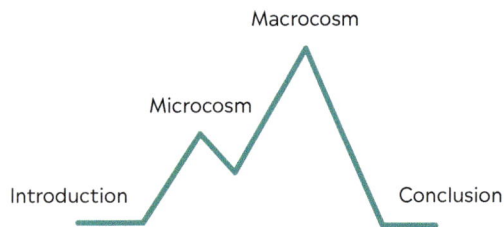

There can be all sorts of variations on this theme. Stories can present an overview first, then zero in on one or more case studies. Or they might alternate from macrocosm to microcosm:

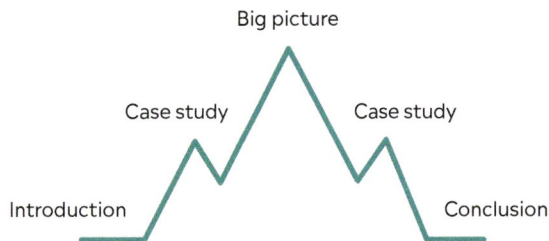

It's useful to think about the overall structure of your story as a way of ensuring that you're taking your readers on a journey that presents your content in a logical, elegant sequence with a clear sense of narrative and progress and with a conclusion that includes some sort of call to action— even if it's simply a link or two to additional reading.

Another way to think about story structure comes to us through Pixar Animation Studios, admired for its brilliant storytelling, a reputation earned from its succession of animated films. Although the "story spine" was popularized by Pixar, it didn't originate with them. Its origin is obscure, but its utility is unquestioned. Here's the spine, with my notes about how they might apply to ArcGIS StoryMaps:

**Once upon a time there was ____.** This part of the story sets the scene and establishes the characters. The scene is especially important in map-driven stories. It's at this point in your story that a locator map is probably appropriate.

**Every day, ____.** Here you can make readers familiar with the world, or the corner of the world, that you're dealing with. Photographs might be appropriate here. They complement maps by depicting how a place looks on the ground.

**One day, ____.** Here's where the drama enters. Is there a threat of some sort? Is the place undergoing change? Has a disaster occurred? Or a policy been implemented?

**Because of that, ____.** What are the consequences of the threat or change?

**Because of that, ____.** There are probably additional consequences; maybe they're cumulative.

**Until finally ____.** The climax! Or the solution! Or perhaps we're simply at the end of a temporal sequence.

**And ever since that day, ____.** This denouement might be a lesson learned, a new practice applied, a predictive model interpreted.

This sequence applies to countless fables, novels, short stories, and screenplays. But it feels less than universally relevant as a formula for location-based multimedia storytelling.

Regardless of the peaks, curves, valleys, and plot points that form the middle of your stories, their start and finish are particularly important. Your story's beginning performs some utilitarian functions, including branding and (sometimes) a byline and date. But its most important job is to lure readers into your narrative with a title that piques readers' curiosity, a subtitle that provides a bit more information about the narrative to follow, and, usually, a compelling and evocative image.

A vital function of the end of your story—that I'll return to later in the book—is to give your readers some sort of *call to action*: buttons for volunteering or donating, places to go for more information, links to additional reading, click-throughs to organizations, and so on. Ideally, it will also feature a concluding paragraph or two—not unlike that "and ever since that day" piece that forms the final vertebra of the story spine.

Your story's middle, regardless of the structure it follows, should inform, inspire, provoke. It should *change minds*, by describing a place, presenting plans, predicting the future, rallying readers to a cause, making a piece of the world—no matter how small or how remote—more understandable. It should engage your readers, luring them step by step through elements that are clearly presented, flow smoothly from one item to the next, and—in the best of circumstances—elicit readers' empathy and inspiration.

## Maps and stories

Although the general principles I've touched on apply to stories of all sorts, our focus in this book is mainly on the distinctive roles that maps play as storytelling devices.

Maps in storytelling contexts anchor narratives in a location; they guide us from place to place. They show change over time; they reveal the past and predict the future; they parse geographies into components and categories that reveal patterns and interrelationships. Maps add depth and dimension to stories that are difficult or impossible to achieve with other media.

In chapter 2, we'll compare and contrast how words, pictures, and maps provide contrasting means of answering the reporting and storytelling basics of *who, what, where, when, why,* and *how.* In chapter 6, we'll consider the various roles maps play in storytelling contexts, from bit parts to stars of the show. Although we'll occasionally reminisce about maps in the analog age, we'll primarily discuss modern, digital storytelling, citing most frequently Esri's popular storytelling app, ArcGIS StoryMaps.

## What is ArcGIS StoryMaps?

ArcGIS StoryMaps enables the creation of web-based narratives that combine maps and other multimedia elements—words, images, videos, audio, embedded content—to tell stories about the world.

ArcGIS StoryMaps has democratized multimedia storytelling. Its intuitive builder functions enable storytellers to create elegant multimedia narratives with little or no instruction. No web development or programming skills are needed to create stories. A single individual can create professional-quality narratives that used to require teams of developers, designers, cartographers, and editors. Not required: years of technical training. Required: creative inspiration and a modicum of editorial skills.

Stories created with ArcGIS StoryMaps can cover almost any topic. They can tell stories at all scales, from neighborhoods to cities, regions to continents, oceans to the whole world. Their primary components (*facing page*) include **(1)** a cover, usually featuring a title, subtitle, and image;

1

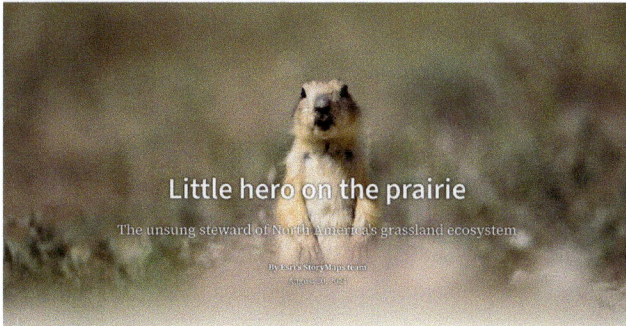

2

## Friend or foe?

Prairie dogs have historically been deprived of the accolades they deserve. In fact, for much of history, they have been poisoned, hunted, and regarded as pests. It is still common for recreational shooters to use prairie dogs for target practice. Many landowners believe prairie dog tunnels destroy property and ruin grazing lands by eating the grasses meant for cattle. Despite these beliefs, the competition between cattle and prairie dogs is still under-researched and hotly contested. In addition to human threats, prairie dogs are at risk of disease; fleas carrying sylvatic plague can wipe out entire colonies.

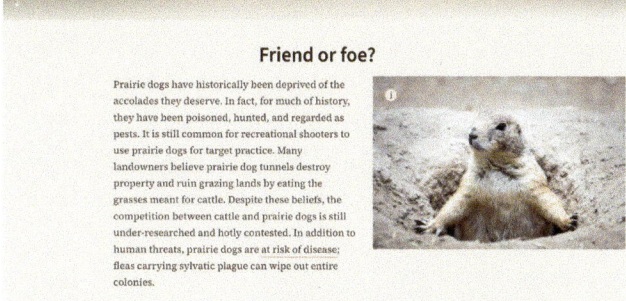

3

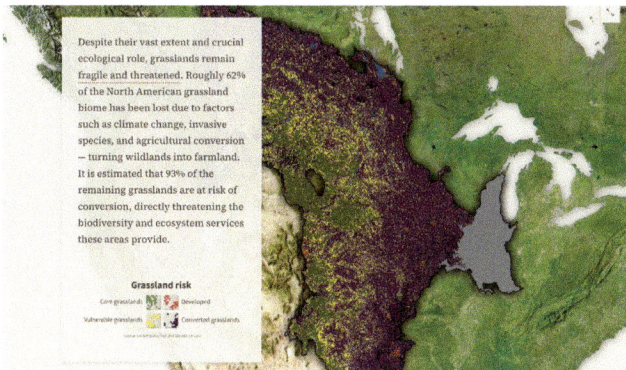

Despite their vast extent and crucial ecological role, grasslands remain fragile and threatened. Roughly 62% of the North American grassland biome has been lost due to factors such as climate change, invasive species, and agricultural conversion — turning wildlands into farmland. It is estimated that 93% of the remaining grasslands are at risk of conversion, directly threatening the biodiversity and ecosystem services these areas provide.

4

**The Prairie Dog Project**

Showcasing 44 years of prairie dog research by behavior ecologist John Hoogland.

https://www.prairiedoghoogland.com

**Prairie Dog Coalition**

The Prairie Dog Coalition (PDC) is an alliance of non-profits, scientists, advocates, and concerned citizens.

https://www.prairiedogcoalition.org

5

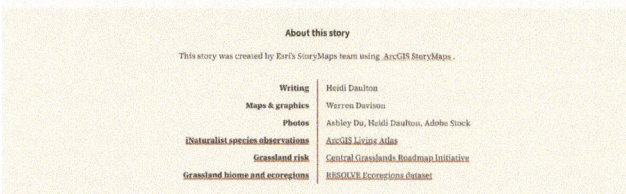

**About this story**

This story was created by Esri's StoryMaps team using ArcGIS StoryMaps .

| | |
|---|---|
| Writing | Heidi Daulton |
| Maps & graphics | Warren Davison |
| Photos | Ashley Du, Heidi Daulton, Adobe Stock |
| iNaturalist species observations | ArcGIS Living Atlas |
| Grassland risk | Central Grasslands Roadmap Initiative |
| Grassland biome and ecoregions | RESOLVE Ecoregions dataset |

1

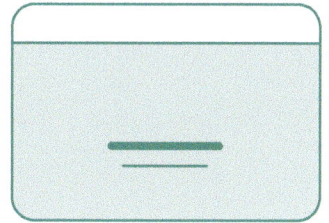

2

3

4

5

**Little Hero on the Prairie**
⊂⊃, an ArcGIS StoryMaps story created by the Esri ArcGIS StoryMaps team.

**(2)** a scrolling narrative format that accommodates long-form text and other multimedia items; and **(3)** "immersive" sections that enable a variety of storytelling formats and approaches, including tours and slideshows. Stories can include **(4)** embedded content in the form of cards or interactive windows, and they feature **(5)** a concluding credits section.

In addition to scrolling narratives, ArcGIS StoryMaps has three additional outputs: briefings, which replace the vertical scrolling action with a horizontal, slide-based experience for in-person presentations; collections, which facilitate groupings of stories and other items into several gallery formats; and themes, which allow authors to customize colors, fonts, and other elements.

Thousands of individuals and organizations have found ArcGIS Story-Maps features and functions to be useful:

It's **interactive**. Many organizations that customarily publish articles and reports online as static PDF files find the interactivity and visual richness of stories created with ArcGIS StoryMaps to be more engaging to their audiences. Stories can also incorporate dynamically updated web maps and scenes, reducing the likelihood that stories will become dated.

It's **flexible**. Features such as immersive sections and custom themes enable authors to create stories that vary greatly in length, structure, topic, and appearance. Stories can be duplicated within and across organizations; authors can produce and share preformatted templates that classrooms and communities can use to create multiple stories with similar structures and topics.

Stories are **hosted** by Esri on its cloud service, ArcGIS Online. Story authors needn't worry about maintaining servers and other infrastructure. Hosting also ensures that stories benefit from **frequent product updates**, many of which are developed in response to feature requests from the storytelling community.

Stories are **accessible** and **responsive**. Authors can create alternative text that is detectable by screen readers; color pickers alert authors when contrast between type and background hues is problematic for color-blind readers. Stories display well on a variety of screen sizes and proportions, ensuring satisfactory experiences on PCs, tablets, and smartphones alike.

ArcGIS StoryMaps can access **geospatial content** in the form of web maps, layers, and 3D scenes created by organizations and shared in ArcGIS Online, including curated, authoritative content featured in ArcGIS Living Atlas of the World.

Stories created with ArcGIS StoryMaps are **sticky**; in other words, readers on average spend far longer, and read deeper into, multimedia stories than they typically do on standard web pages.

They enable geographic information system (GIS) professionals to **tell the story of their work**. GIS users can interpret and publish their maps and analyses, making their insights available to people within organizations, to professional networks, and to the public at large.

For students and educators, ArcGIS StoryMaps is a **teaching and learning tool**. Instructional stories bring topics in the sciences and humanities to life. Thousands of students are creating their own multimedia stories, often as alternatives to traditional (and static) research papers. Stories have become a gateway to geography for many students who might otherwise never have encountered GIS.

You'll find much more detail about ArcGIS StoryMaps in chapter 8.

## The uses of ArcGIS StoryMaps

ArcGIS StoryMaps has been put to use in myriad ways. We've seen a lot of résumés in ArcGIS StoryMaps form, but we've also found stories guiding guests to wedding ceremony and reception venues. Categorizing narratives is a daunting task, considering the number of stories (over three million Esri-hosted ArcGIS StoryMaps stories alone) and the degree to which stories commonly defy easy pigeonholing because of their mix of purposes, topics, and components.

After perusing thousands of stories, and after creating scores of them over the past dozen years, I've come up with a list of 15 story types that I've labeled as verbs. I've further grouped them into five larger categories:

| Understand | Educate | Explore | Document | Entertain |
|---|---|---|---|---|
| Interpret | Teach | Guide | Collect | Share |
| Document | Learn | Present | Curate | Wander |
| Compare | Instruct | Recall | Report | |
| Inspire | | | | |

Many stories fit into two or more of these categories, but I'm hopeful that listing and describing this suggested taxonomy will reveal some patterns and perhaps provide some inspiration. On the following pages, I briefly describe each category and subcategory and include an example or two of each.

## Understand

These stories inform people about issues and phenomena.

## *Interpret*

Maps combine with text, graphics, and images to make geographic data understandable to broad audiences.

Example story: **America's Mental Health Crisis, Mapped** ⊖⊃, by Esri chief medical officer Este Geraghty, summarizes in maps, text, and graphics an unprecedented US medical health crisis and includes comparative maps of adult obesity, frequent mental distress, unemployment, and other factors. Mapped here: average number of mental health days in 2015 and 2019.

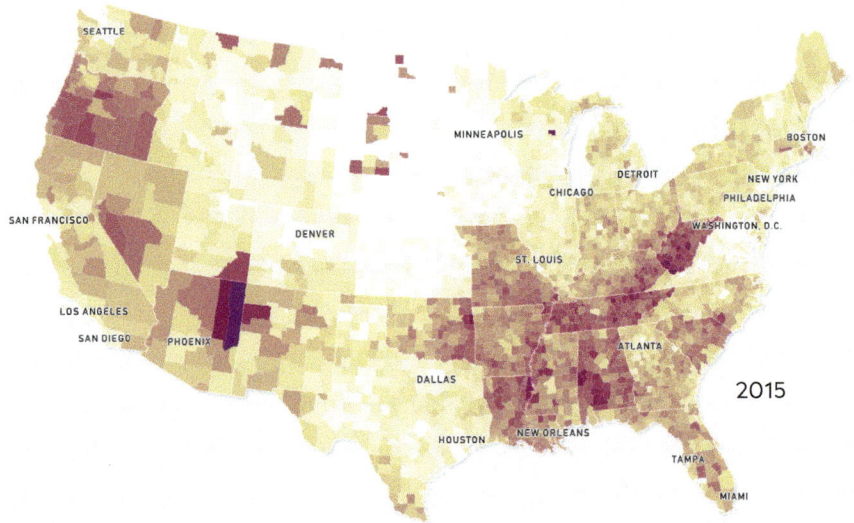

2015

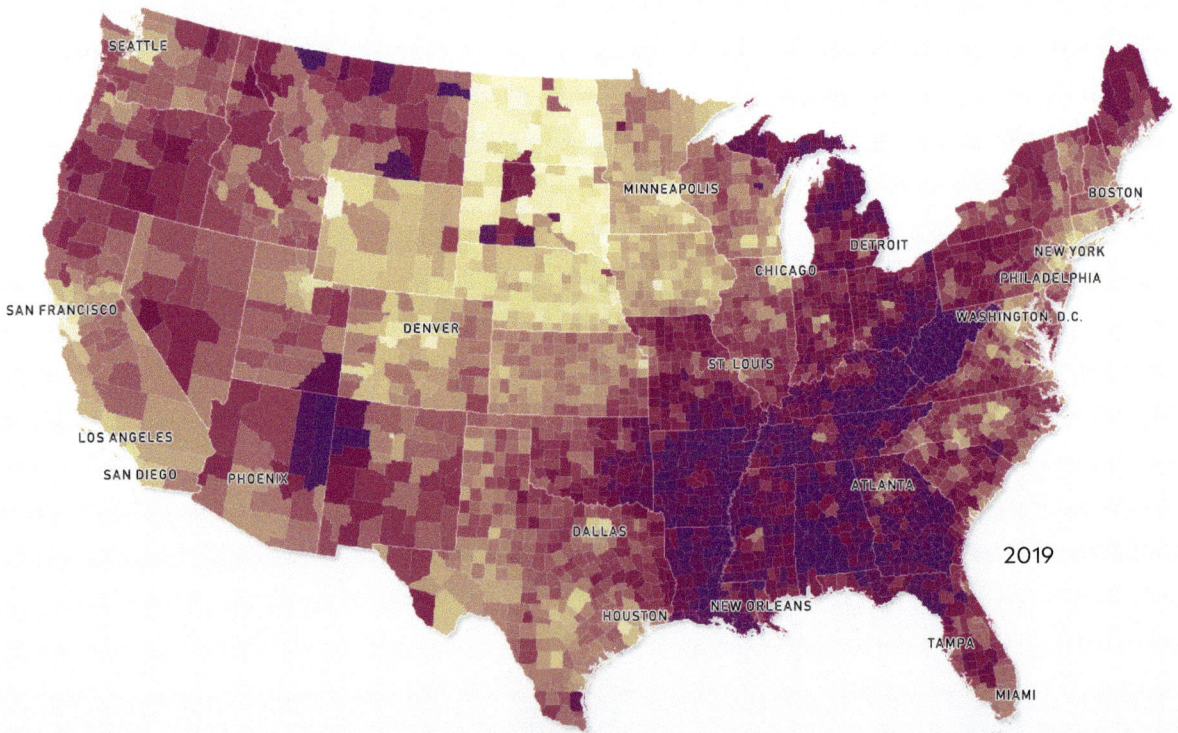

2019

## Document

Many ArcGIS StoryMaps stories present and interpret GIS analyses.

Alcis, a firm that specializes in GIS work on behalf of underserved communities, collaborated with Seeds for Development, a UK-based nonprofit, in an effort to improve food security in areas of northern Kenya. Its **Satellites and Seeds** ⊖ story describes how they mapped hundreds of roads and dwellings, located schools that could serve as potential distribution hubs, and estimated travel times. Photos and videos humanize the story.

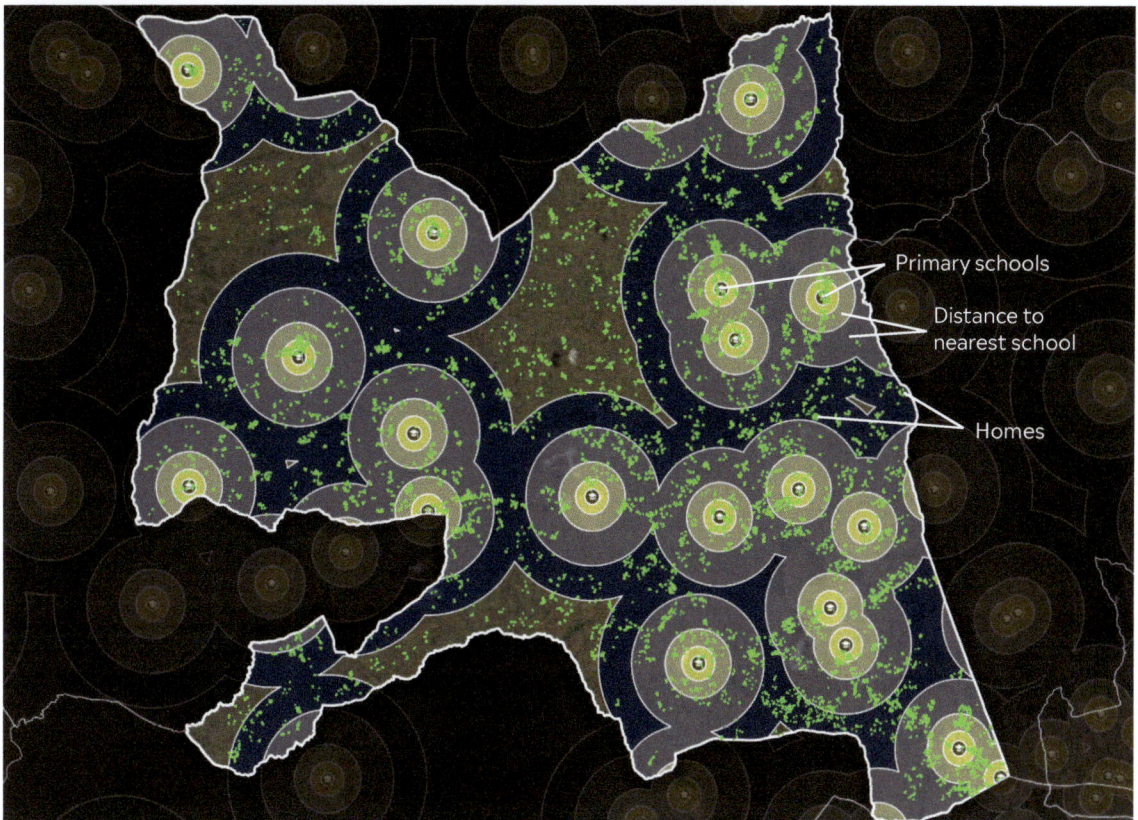

Primary schools

Distance to nearest school

Homes

## *Compare*

Some stories reveal the complex textures of a special place; others compare and contrast multiple locations.

> **Thirteen Ways of Looking at the Grand Canyon** ⟨→⟩ explores multiple facets of the iconic national park by displaying a series of thematic maps at identical scales.

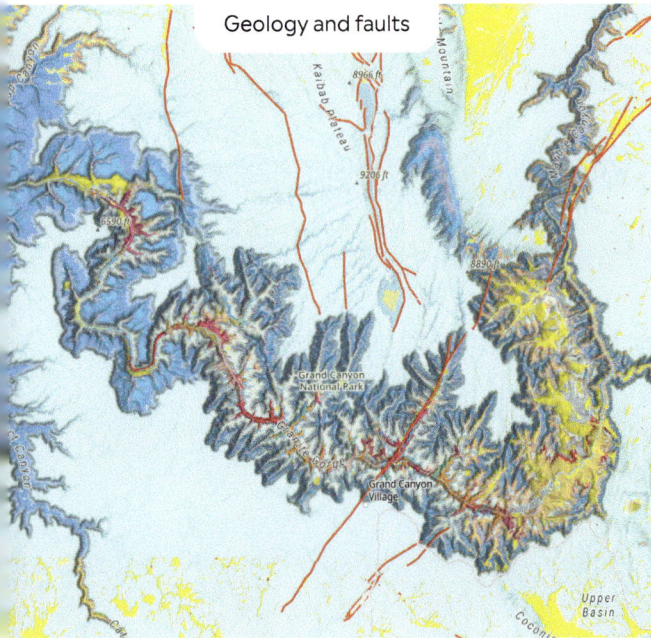
Geology and faults

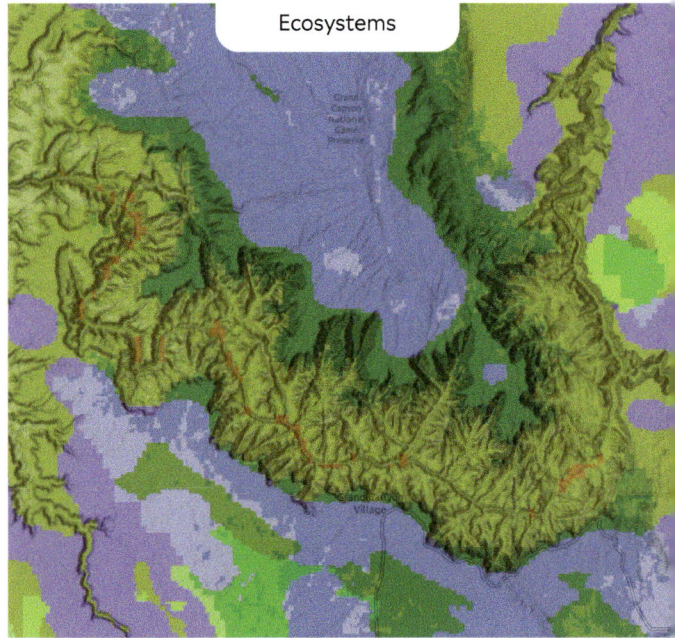
Ecosystems

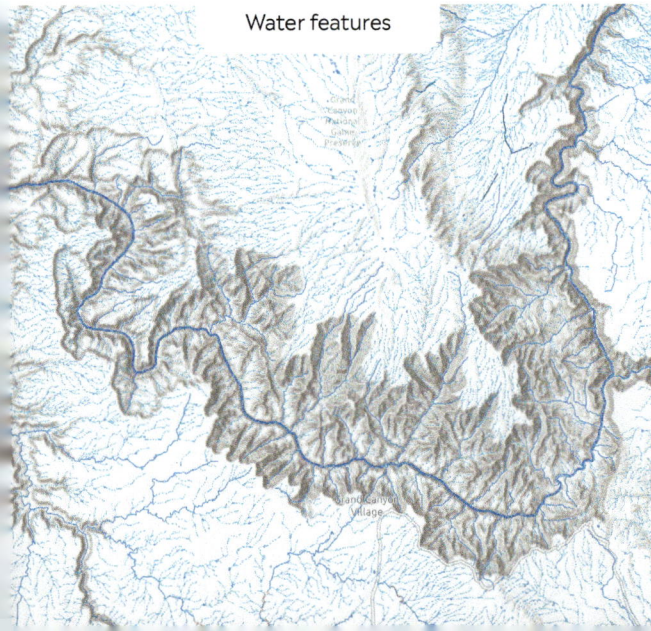
Water features

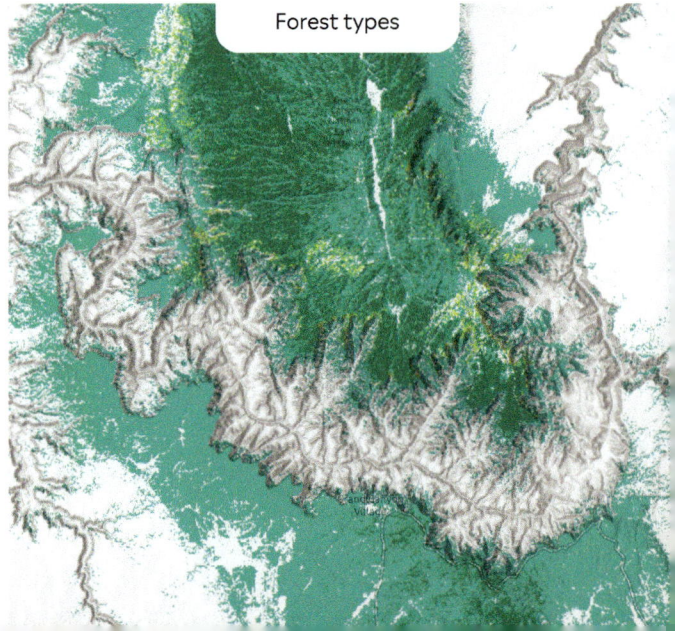
Forest types

## *Inspire*

Inspirational stories open people's eyes to the world by vividly portraying distant places and diverse cultures.

The Amazon Conservation Team's **Living Territories** ⟨⊖⟩ story melds cartography, artwork, photographs, and text to make an inspiring case for protection of Indigenous groups and biological diversity in the Amazon basin. Skillfully choreographed maps portray Indigenous lands, sacred landscapes, and increasing threats from mining and development.

Bogotá

COLOMBIA

### Guardians of the Sacred Headwaters

In **southern Colombia**, expanded territorial rights have consolidated protection of a **biocultural conservation corridor** across the **Andes-Amazon** transitional region, ensuring Indigenous stewardship of threatened ecosystems critical for **medicinal plants** and **water resources** in a mountainous region where several significant Amazon tributaries are born.

## Educate

Thousands of narratives inform people about issues and phenomena.

## Teach

These stories provide multimedia narratives for classroom instruction or informal education.

**Living in the Age of Humans** ⊖, by the Esri ArcGIS StoryMaps team, is a series of visually engaging stories depicting the profound and increasing impact of human activities on the natural world. The stories have perennially ranked among the most visited of the team's stories. *Below*: The world biomes. *Right*: Terrestrial ecosystems of the world.

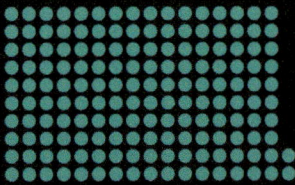
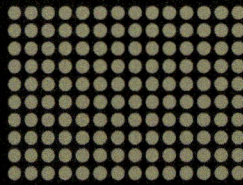
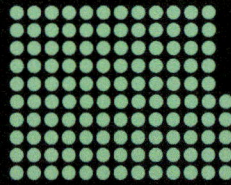
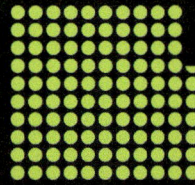

Tree cover    Bare areas    Cropland and pasture    Rangeland and some natural grassland    Sparse areas    Water, marsh    Urban areas

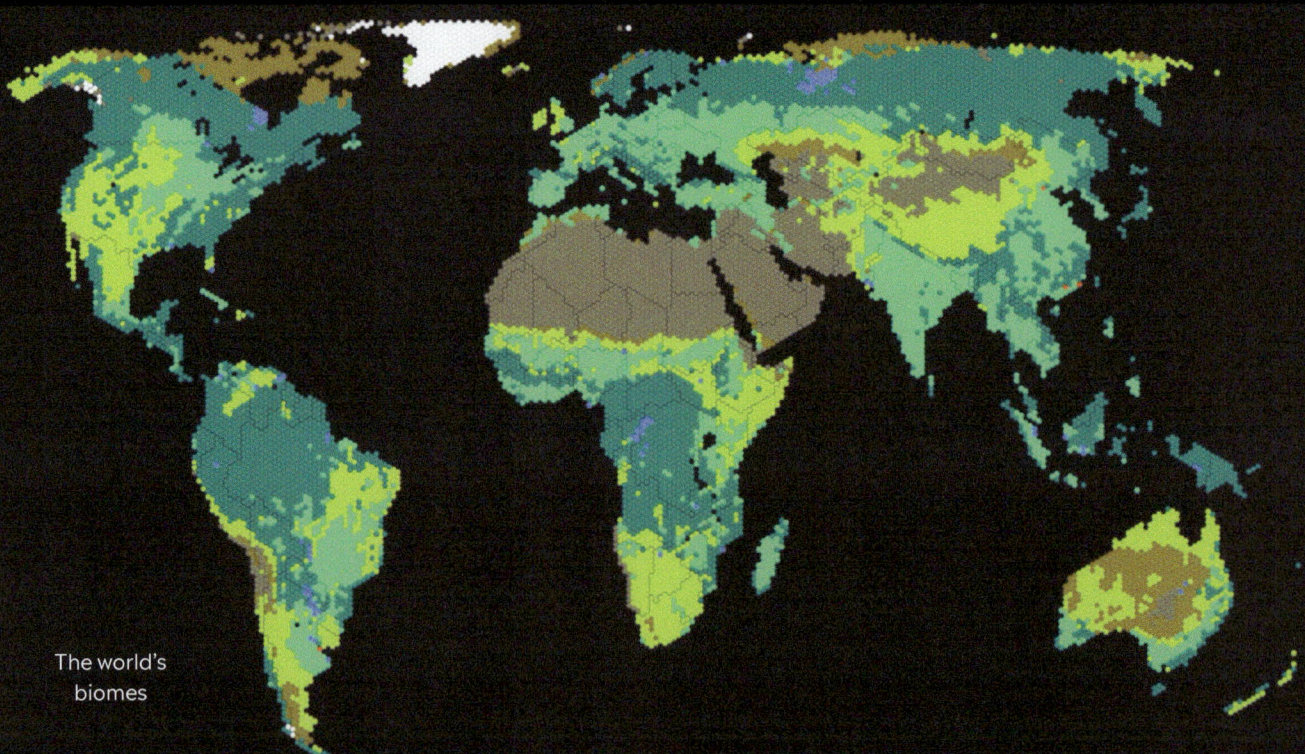

The world's biomes

## *Learn*

ArcGIS StoryMaps empowers students to create their own stories and exposes them to the insights of the geographic approach.

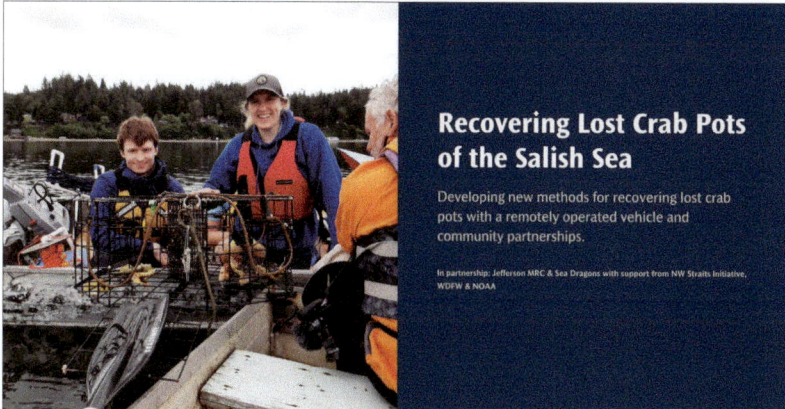

### Recovering Lost Crab Pots of the Salish Sea

Developing new methods for recovering lost crab pots with a remotely operated vehicle and community partnerships.

In partnership: Jefferson MRC & Sea Dragons with support from NW Straits Initiative, WDFW & NOAA

### Meet the Team

The Jefferson MRC is part of the Northwest Straits Marine Conservation Initiative (which includes NW Straits Commission, NW Straits Foundation, and seven local Marine Resources Committees) and serves as an advisory group to the Jefferson Board of County Commissioners. The MRC also engages in hands-on marine stewardship projects with community partners, such as the Sea Dragons, to restore and protect the marine and nearshore environments of East Jefferson County.

2022 team (left to right): Jeff Taylor, Logan Flanagan, Riley Forth, Nathaniel Ashford, Ella Ashford, Monica Montgomery.

"The MRC, with support and guidance from the NW Straits Foundation derelict gear removal program, secures permits and approval from the WA Department of Fish and Wildlife, ensures adherence to the rules and regulations, contracts side scan sonar surveys, and coordinates fieldwork" - Monica Montgomery (MRC Coordinator)

### Side-Scan Sonar Findings

The side-scan sonar data was thoroughly filtered in order to identify the features most likely to be a crab pot. Sonar surveying revealed many possible crab pots in Port Townsend Bay south of the ferry line. The team concentrated their efforts in this region since it was safer to operate outside of the ferry path and the area had a high prevalence of marine debris. Once the ROV dives commenced, there were several surprising discoveries.

Preparing to launch the ROV in Port Townsend Bay

Students by the thousands are creating multimedia narratives, many of them as alternatives to traditional, static research papers. For many students, stories are an initial gateway to deeper involvement with geography and GIS. **Recovering Lost Crab Pots of the Salish Sea** ⟲ was a winner in the student category of Esri's 2023 Storytelling with Maps competition.

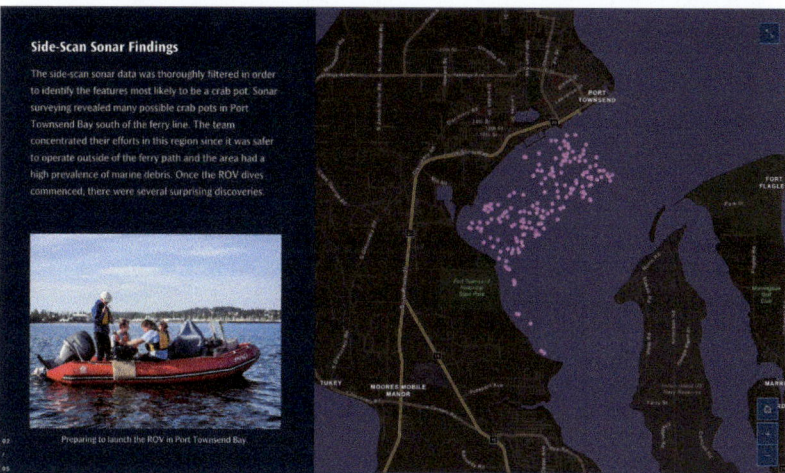

## *Instruct*

Step-by-step tutorials benefit from interactivity.

**Combining Crowdsourced Data and ArcGIS StoryMaps** ⊖ is an example of a technical tutorial that benefits from the multimedia capabilities of ArcGIS StoryMaps narratives. Screen recordings of survey app functions, map editing, and publishing processes reinforce and complement text descriptions.

### Creating a Survey:

Survey123 is a web application within the Esri ecosystem used to generate forms and catalog responses. If you're getting started with Survey123 try watching this short video.

A useful component of Survey123 is the **Map** question field, which allows readers to submit location information as part of the survey. Be sure to include this question if you want to map the survey results in your story.

It's best practice to always add a question asking responders to opt into sharing their information with you. Clearly communicate how the submitted data will be used so submitters can decide if they're comfortable with the end use of the data. If the data is being used for commercial purposes, you should indicate this in the term of use question.

Data transparency is essential, and you should seek legal advice if you're collecting sensitive data before making your survey public.

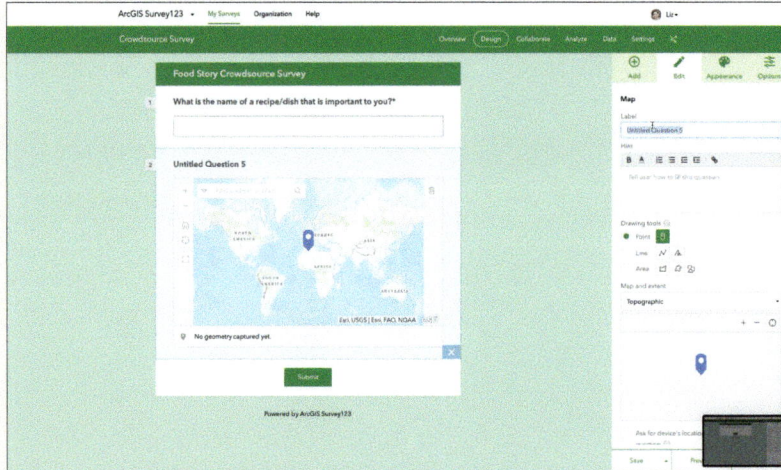

### Publishing your survey

Once you've created the questions for your survey click **Publish**. You'll need to update your sharing settings to enable the public to submit responses to the survey. Test the survey and submit a point or two. Seeding your crowdsourced effort with a few sample submissions will encourage others to join it and provide examples of the type of submissions you're after.

### Map your survey results

You have two easy ways to display your survey results: You can add them to your story using the map block or incorporate them directly through a data driven map tour in ArcGIS StoryMaps.

Let's walk through each of these options.

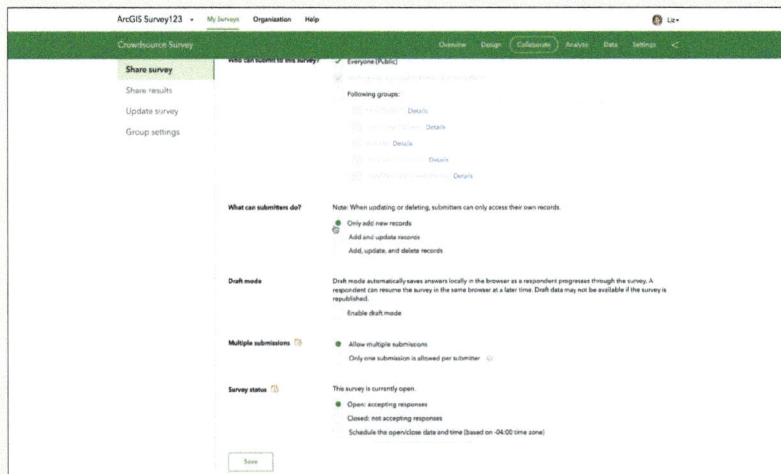

## Explore

Many stories inform people about issues and phenomena.

## *Guide*

Stories take readers from place to place—within a locale or across the globe.

In **Endemic Species** ⊝, Esri's ArcGIS StoryMaps team gave a light touch to a weighty topic, profiling 10 species across the globe that inhabit limited areas and are threatened by human activities. Team member Warren Davison created animated, collage-style illustrations that introduce readers to the species and their habitats.

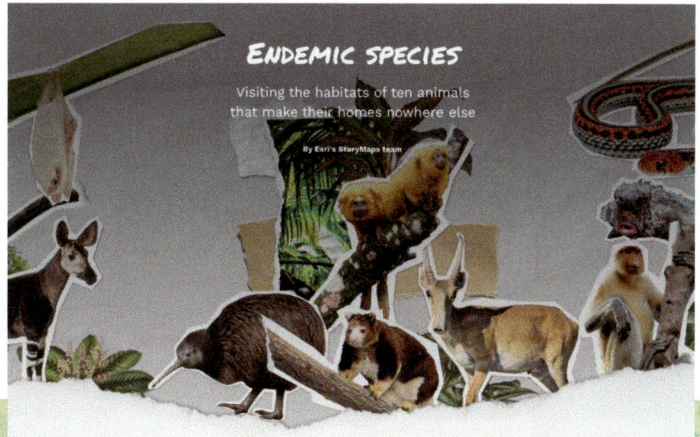

ENDEMIC SPECIES

Visiting the habitats of ten animals that make their homes nowhere else

By Esri's StoryMaps team

## OKAPI

Is it a giraffe? A horse? A zebra? Well, if you're in the northeastern portion of the Democratic Republic of the Congo, it's likely an okapi! This species is also called a "forest giraffe" and resembles a deer in form-fitting, striped pants. Okapis live under forest canopies and are solitary animals, which is a big surprise considering how cool they look.

## *Present*

Briefings enable in-person, location-enabled presentations.

**Residential Redevelopment in the West Don Lands** ⊝, by the Esri ArcGIS StoryMaps team, depicts a large development east of downtown Toronto, Canada. It demonstrates the capabilities of ArcGIS StoryMaps' slide-based briefings output. In-person presentations benefit from interactive maps and web scenes and can be viewed offline in a tablet app.

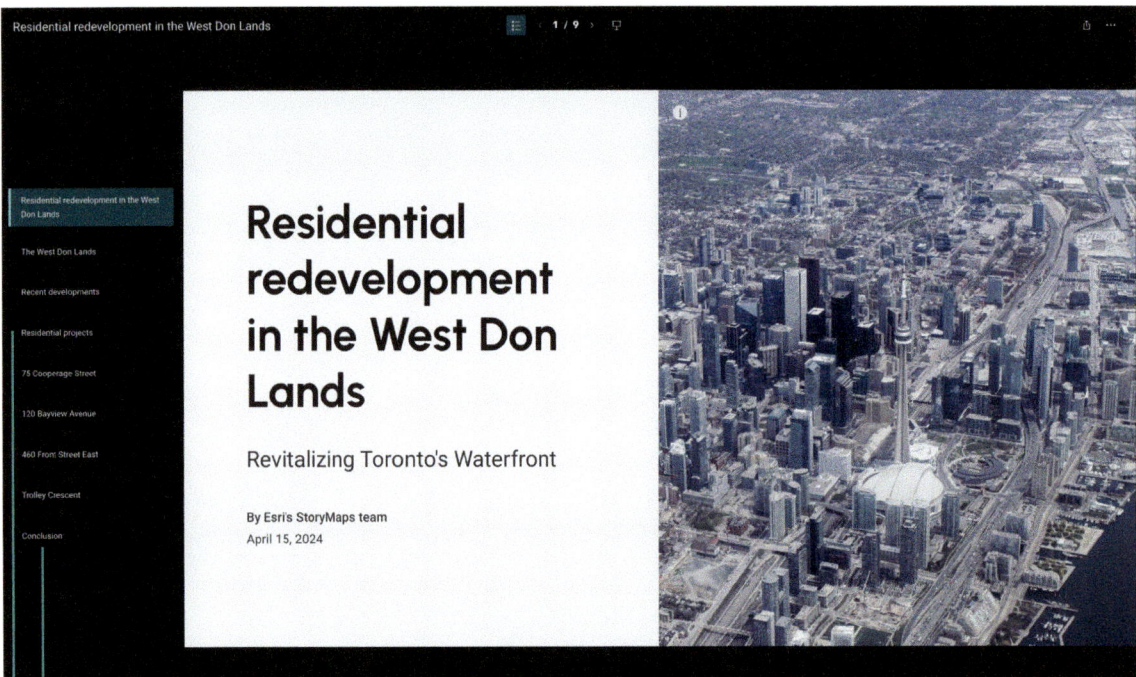

Residential redevelopment in the West Don Lands          1 / 9

# Residential redevelopment in the West Don Lands

Revitalizing Toronto's Waterfront

By Esri's StoryMaps team
April 15, 2024

Sidebar navigation:
- Residential redevelopment in the West Don Lands
- The West Don Lands
- Recent developments
- Residential projects
- 75 Cooperage Street
- 120 Bayview Avenue
- 460 Front Street East
- Trolley Crescent
- Conclusion

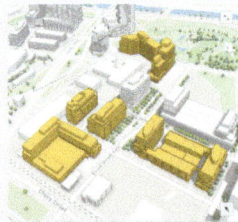

**Residential projects**

Many of these newly constructed buildings are either residential or mixed-use and add nearly **6,000 residential units** to the neighborhood.

**Conclusion**

When completed, the West Don Lands redevelopment project will contain:

- Around 6,000 new residential units.
- An elementary school.
- Daycare centers.
- A plethora of new commercial and retail space.
- Over 20 acres of park areas.

All of these amenities are contributing to the revitalization of Toronto's Waterfront.

⊟ Read the full story here.

## *Recall*

History comes to life with maps and archival multimedia.

MRS. FRAZER BAKER AND CHILDREN.

### Tulsa Star

The *Star* was an influential newspaper that championed Black causes, promoting progress and stability within Tulsa's Black community.

**The Burning of Greenwood** ⬡, by National Geographic Education, recounts the horrific 1921 massacre in Tulsa, Oklahoma, that destroyed the thriving majority-Black neighborhood of Greenwood.

### Frisco Railroad Tracks

In Tulsa, laws prevented both White and Black people from living in

## Document

Stories inform people about issues and phenomena.

## Collect

Collections can aggregate stories or present stories as e-book chapters.

**Storytelling for a Sustainable World** ⊖, created by the Esri ArcGIS StoryMaps team, features one multimedia narrative for each of the 17 United Nations Sustainable Development Goals. The stories share a common design and narrative structure and are aggregated on a single collection page.

Collection

# Storytelling for a Sustainable World

This collection, which includes one story for each United Nations Sustainable Development Goal, provides sample stories, data sets and applications, and learning resources for getting started with place-based storytelling.

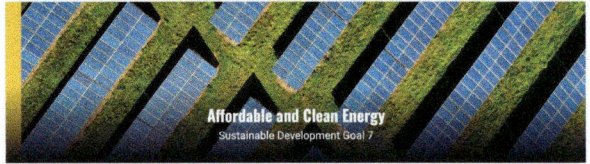

**Storytelling for a Sustainable World**
Advance progress towards the Sustainable Development Goals with data-driven storytelling

**Clean Water and Sanitation**
Sustainable Development Goal 6

**No Poverty**
Sustainable Development Goal 1

**Affordable and Clean Energy**
Sustainable Development Goal 7

**Zero Hunger**
Sustainable Development Goal 2

**Decent Work and Economic Growth**
Sustainable Development Goal 8

**Good Health and Well-Being**
Sustainable Development Goal 3

**Industry, Innovation, and Infrastructure**
Sustainable Development Goal 9

**Quality Education**
Sustainable Development Goal 4

**Reduced Inequalities**
Sustainable Development Goal 10

## Curate

Web-based stories make museum and zoo exhibitions available to audiences unable to visit in person.

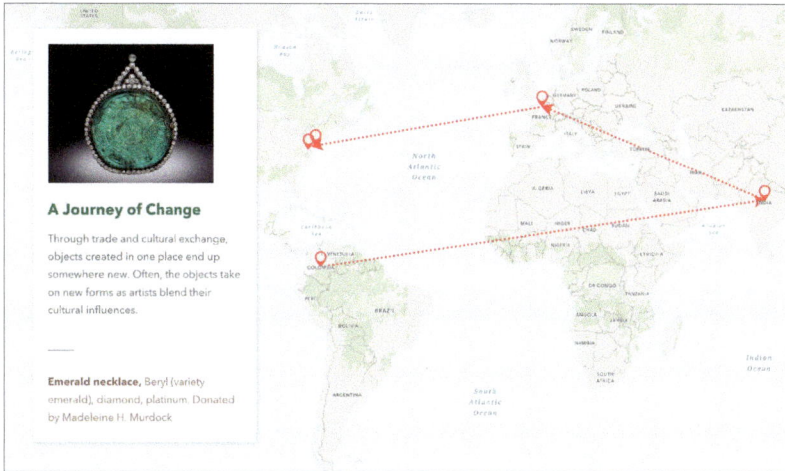

**A Journey of Change**

Through trade and cultural exchange, objects created in one place end up somewhere new. Often, the objects take on new forms as artists blend their cultural influences.

**Emerald necklace,** Beryl (variety emerald), diamond, platinum. Donated by Madeleine H. Murdock

**Objects of Wonder** ⊂⊃, by the Smithsonian's National Museum of Natural History, is a digital companion to a physical exhibition. It features a wide range of objects from the museum's vast collection, comprising more than 147 million objects.

## *Report*

These stories provide information about the activities of an organization, including annual reports and online executive summaries.

In **Geospatial Conservation at The Nature Conservancy** ⊖⊃, the organization's GIS team focused its annual report on the role of Earth observation in advancing science, policy, and decision-making. The narrative is enriched with maps and aerial imagery.

### The Nature Conservancy

## Global Projects

As technology advances, the growth of EO data from satellites, drones, aircraft and ground-based sensors offers a unique opportunity to measure changes in both natural and human environments consistently. Highlighted here is a small sampling of conservation projects around the world where EO is critical to advancing our 2030 goals.

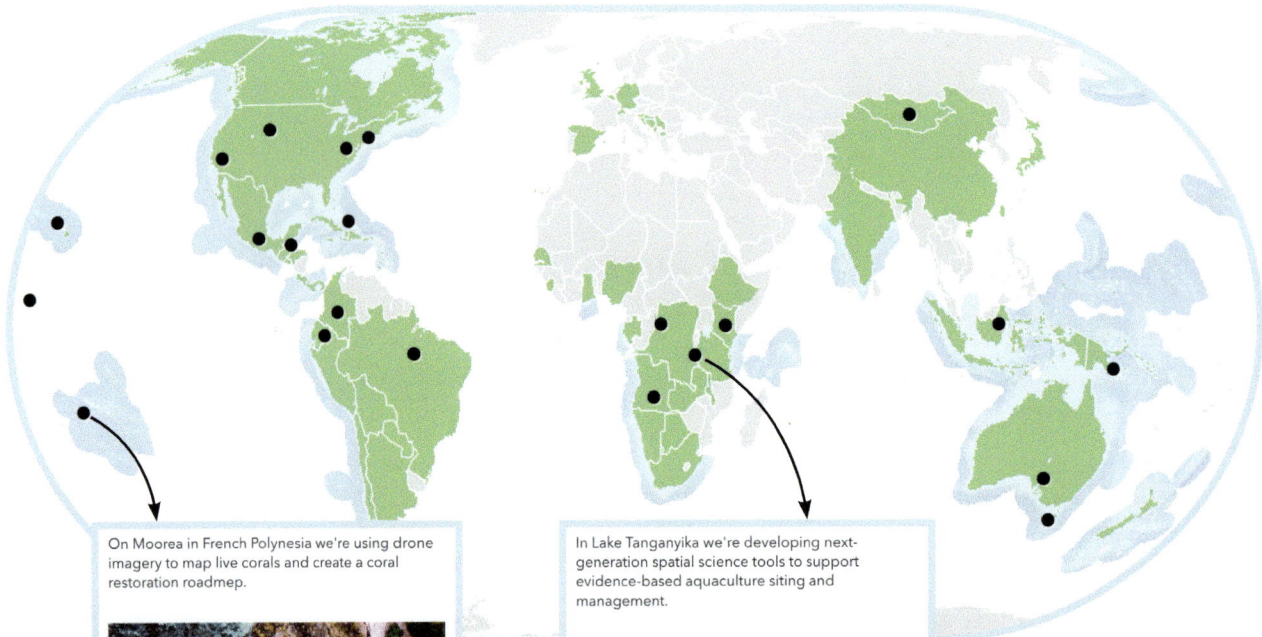

On Moorea in French Polynesia we're using drone imagery to map live corals and create a coral restoration roadmep.

©Vienna Saccomanno/TNC

In Lake Tanganyika we're developing next-generation spatial science tools to support evidence-based aquaculture siting and management.

©Jonathan MacKay/TNC

# Entertain

Stories can entertain and amuse—and maybe sneak in a message.

## *Share*

These stories invite audience participation in community storytelling.

**Show us your natural place***

Use the search bar to find a place by name, or drag the pin to its location.

Find address or place

Galway

Galway

Burren National Park

Earthstar Geographics | Esri, TomTom, Garmin, Foursquare, FAO, METI/NASA, USGS

Powered by Esri

Lat: 53.113534 Lon: -9.695115

**What's the name of this place?***

Inis Mór, Aran Islands, Co. Galway, Ireland

**Share your EarthPlaces** ⟳, by the Esri ArcGIS StoryMaps team, invites readers to upload photos, place-names, and brief descriptions of their favorite places in nature using an embedded survey form. Submissions are automatically added to an interactive globe featuring hundreds of EarthPlaces.

**Inis Mór, Aran Islands, Co. Galway, Ireland**

The most beautiful natural landscape and a fascinating community who speak a unique Irish dialect

## *Wander*

Combining descriptive text with rich visual media creates an ideal stage for recounting and sharing personal adventures.

ArcGIS StoryMaps team member, cartographer, and globetrotter Cooper Thomas recounted his adventures in Kyrgyzstan in **Scraping the Heavens** ⊖→, documenting a guided day hike in Alamedin Gorge in the foothills of the Ala-Too Mountains. Users have shared many travelogues, most of them less exotic than Cooper's.

## Looking Ahead

I hope you've found this collection of verbs helpful. If nothing else, the taxonomy might give you ideas for your own future stories. The companion website for this book, go.esri.com/telling-stories-site, provides links to all the stories I've featured on these pages; you can find many more online in the ArcGIS StoryMaps gallery (doc.arcgis.com/en/arcgis-storymaps /gallery).

In addition to the stories featured here, you'll find multipage spreads between the chapters of this book that describe and illustrate more of my favorite stories. Choosing among thousands of outstanding multimedia narratives is a daunting task. I've no doubt forgotten, or failed to discover, beautiful stories on important topics that deserve a place in the spotlight. I suggest that you search for, and bookmark, websites of individuals and organizations that have authored exemplary stories. Some of the best authors of story narratives publish serially—an indication that others find storytelling as addictive as I do. There's a good chance they'll publish new, and even more distinctive, narratives.

Another source of inspiring stories is Esri's Storytelling with Maps competition; we host it annually and post the winners on our website. A final technique for finding stories is to type "[topic] StoryMap" into Google or another search engine, using a topic of your choice.

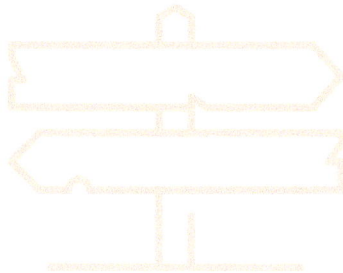

# Living Territories

### Stories of Territorial Justice

**Author:** Amazon Conservation Team

**Medium:** ArcGIS StoryMaps

**Story behind the story:** The Amazon Conservation Team partners with Indigenous and other local communities to protect tropical forests and strengthen traditional cultures. The group has used ArcGIS StoryMaps on multiple occasions to describe the culture, geography, and history of Amazonian rain forest communities—and the environmental and political threats they face—through map-driven digital storytelling.

**Why it's special:** As the story explains, "For Indigenous peoples, territory is more than an ecosystem and place of origin. It is also the source of physical well-being and cultural identity—food, medicine, education, livelihoods, and spiritual practices."

The story seamlessly weaves together maps, artwork, photos, and audio, vividly portraying Colombia's biological and cultural diversity and mapping threats from mining, road-building, and other development pressures.

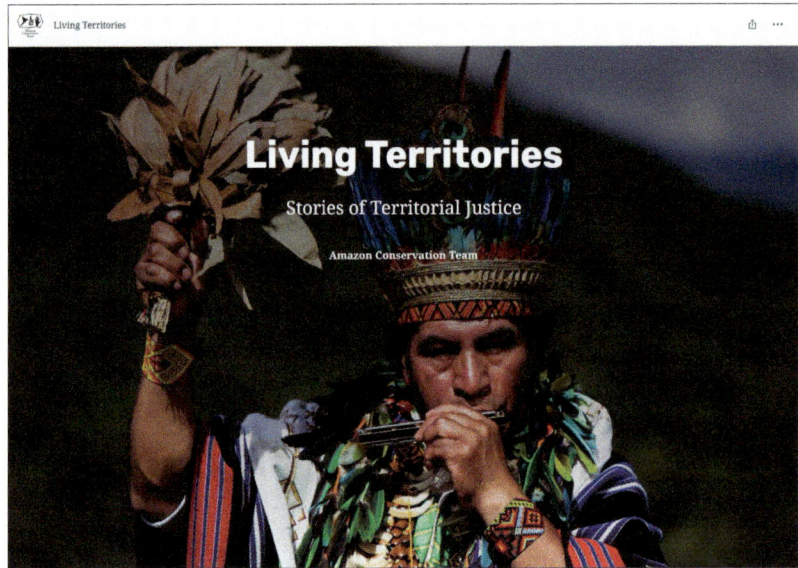

**Living Territories**
Stories of Territorial Justice
Amazon Conservation Team

**Indigenous Land Rights in Colombia**

**Guardians of the Sacred Headwaters**

**Amazon Lowlands & Isolated Indigenous Peoples**

**The Restoration of Sacred Sites and Ancestral Territory**

**Cultural Resiliency in Territory**

**Territorial Claims: Colombia and Regional**

The story is organized into six sections, each of which features a photo essay, map series, and portraits and testimonials from Indigenous people.

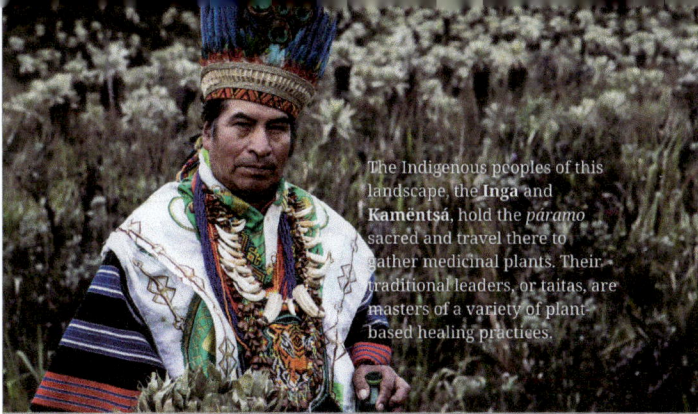

The Indigenous peoples of this landscape, the **Inga** and **Kamëntsá**, hold the *páramo* sacred and travel there to gather medicinal plants. Their traditional leaders, or taitas, are masters of a variety of plant-based healing practices.

Each chapter of the story opens with a photo essay and accompanying text.

## Guardians of the Sacred Headwaters

In **southern Colombia**, expanded territorial rights have consolidated protection of a **biocultural conservation corridor** across the **Andes-Amazon** transitional region, ensuring Indigenous stewardship of threatened ecosystems critical for **medicinal plants** and **water resources** in a mountainous region where several significant Amazon tributaries are born.

Maps and media layers blend art and cartography.

## Mining Concessions and Requests

Antioquia faces severe development pressures, including extensive mining concessions and requests for concessions.

○ Land Titling Area
■ Mining concession (exploitation)
■ Mining concession (exploration)

Source: *Agencia Nacional de Minería*, ANNA platform (accessed September 2023);

There is a concentration of mining activity around El Bagre and Zaragoza in the northeast part of the department, and along the Penderisco River, a tributary of the Atrato River on the western border of the department.

Map choreography locates Indigenous territories and portrays development pressures.

## Voices from the Territory

▷ 00:00 / 00:27 ○————————— ◁))

Listen to Jacinta Jamioy from the Sibundoy Valley

"If you don't have territory, there is no life. We protect the mountains because there is water, the trees because they give us oxygen, and the plains because there, we can plant and grow our food. It is a life process, which serves not only those of us who are in this present moment, but also our future generations. That is why it is a life process."

Jacinta Jamioy, Sibundoy Valley

Portrait photos, text, and audio clips integrate Indigenous voices into the story.

# 1,001 Novels

## A Library of America

**Author:** Susan Straight

**Medium:** ArcGIS StoryMaps

**Story behind the story:** Susan Straight, a novelist based in Riverside, California, grew up with an unusual dual obsession: books and geography. She credits her mother, an immigrant from Switzerland, for her love of books. Her passion for geography came from her stepfather. "We'd go camping in a little trailer. We always had Auto Club maps and a road atlas. I was always obsessed with maps."

**Why it's special:** During the pandemic, Susan began to collect and compile novels by location, listing titles on scraps of paper. Christian Harder, an editor at Esri Press and a neighbor of Susan's, connected her with the ArcGIS StoryMaps team at Esri, and thus began a collaboration that resulted in an online compendium of 1,001 book covers and précis, organized by geography, that was ultimately published by the *Los Angeles Times*.

**The author:** Susan Straight has published 10 books, including *Mecca*, a finalist for the Kirkus Prize in 2022, and *In the Country of Women*, a memoir of migration, books, and family. She was born in Riverside, where she is Distinguished Professor of Creative Writing at UC Riverside.

### The Thousand Crimes of Ming Tsu by Tom Lin (2022)

Corinne, Utah:
A western like no other, with Ming exacting revenge on the men who kidnapped him to work on the railroad, taking him away from his wife, a tale of retribution and history that cannot be put down.

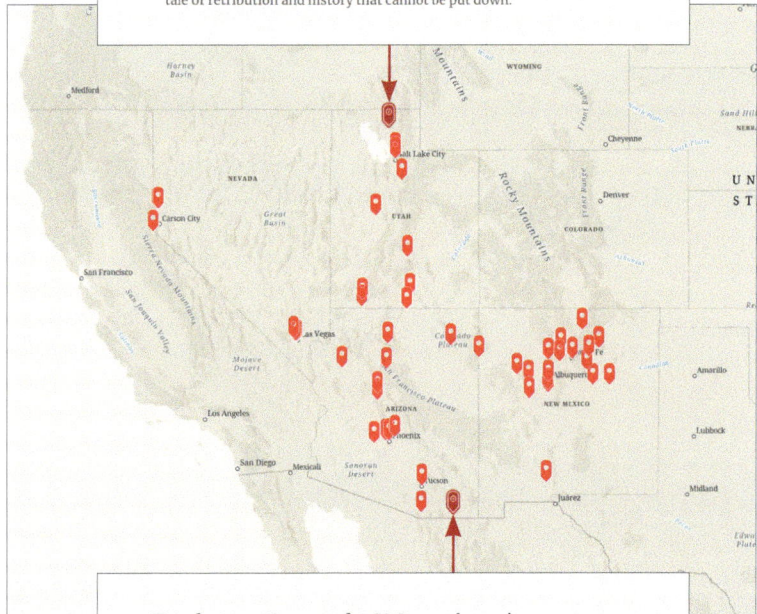

### Tombstone Courage by JA Jance (1994)

Bisbee, Arizona:
Joanna Brady has just been elected sheriff of her hometown, and a deep mine holding two bodies leads her to uncover decades of secrets in this long-running series.

*Above*: Two books featured in the story's Enchanted Deserts & Coyote Canyons section.

*Left*: Susan Straight.

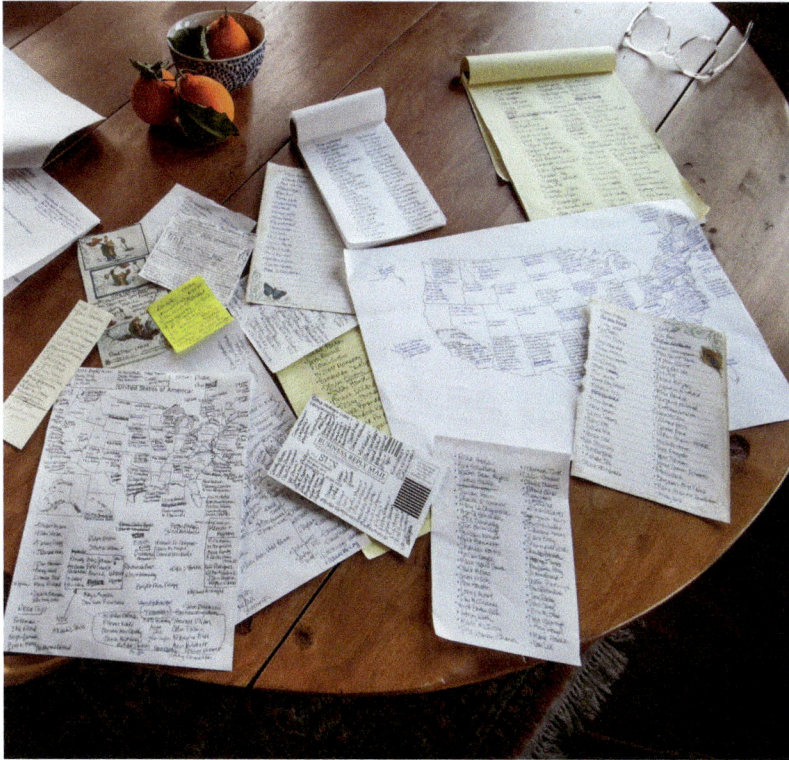

Susan's handwritten book lists, before her effort went digital.

*Below*: An interactive map featuring all 1,001 volumes (not shown: Alaska and Hawaii).

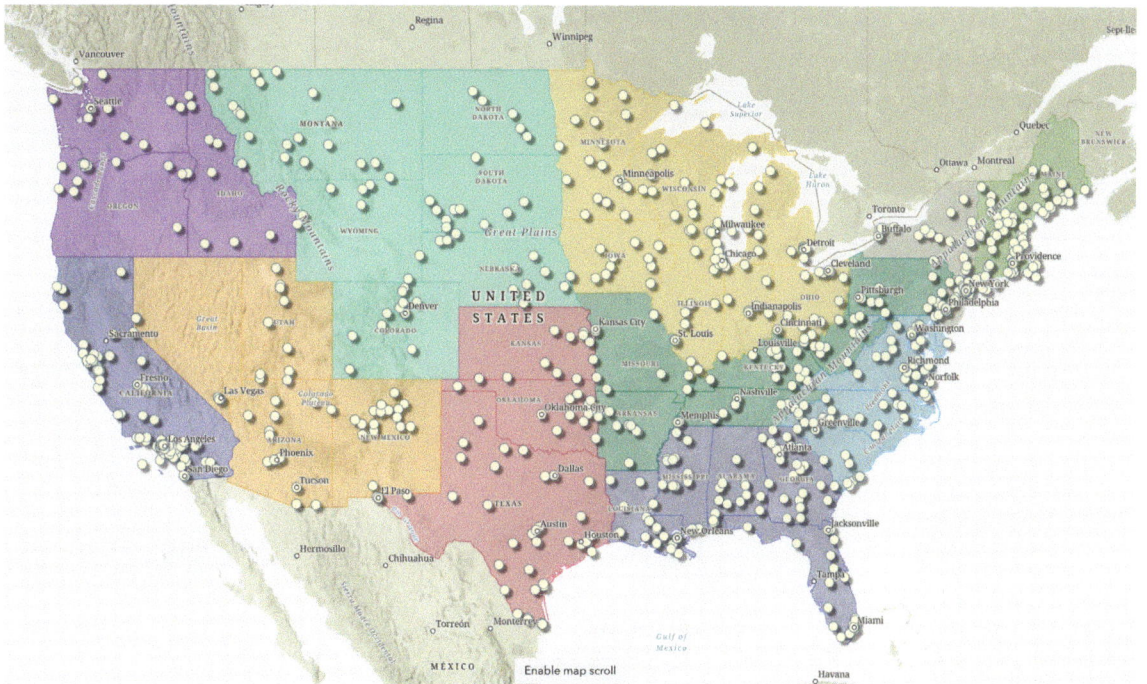

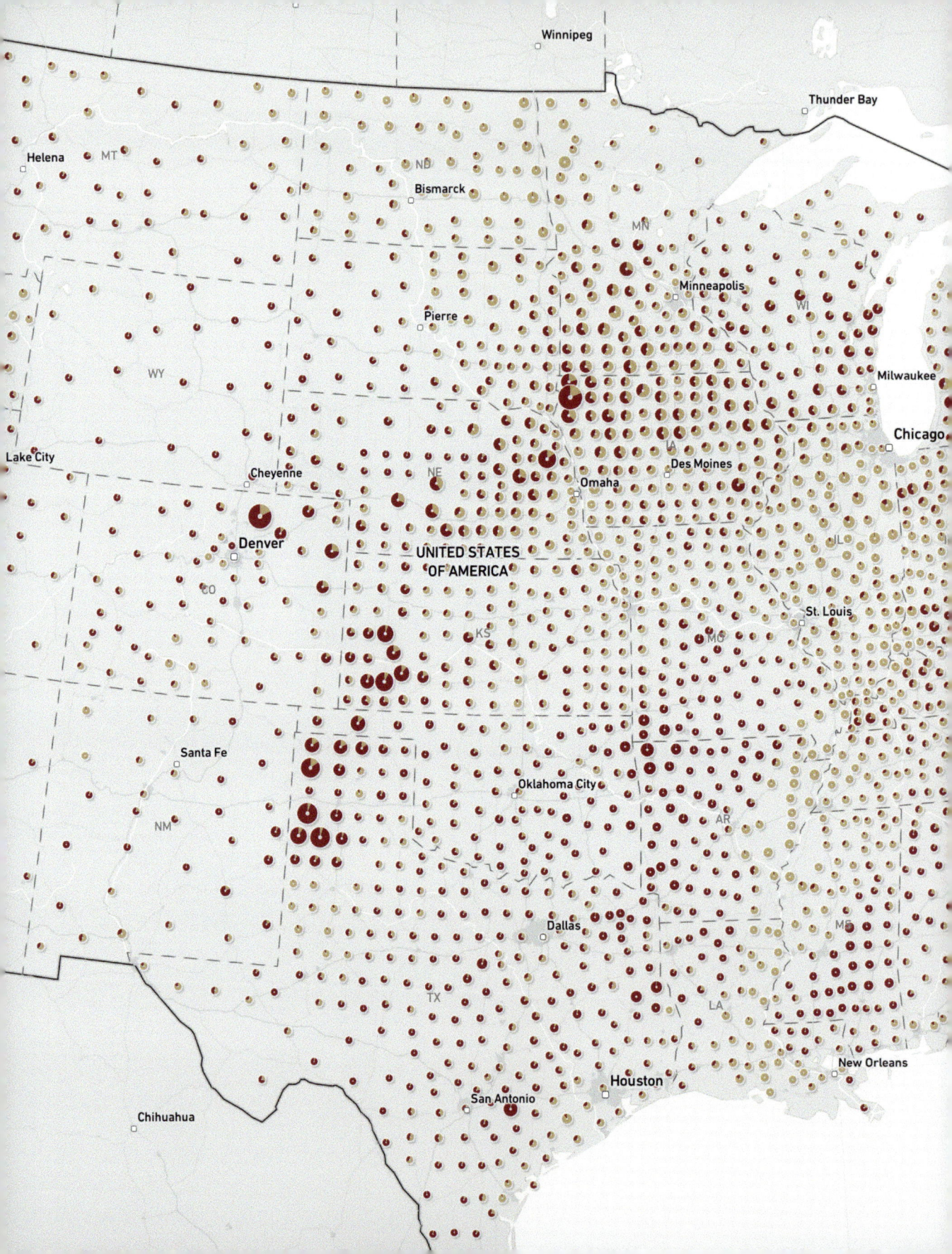

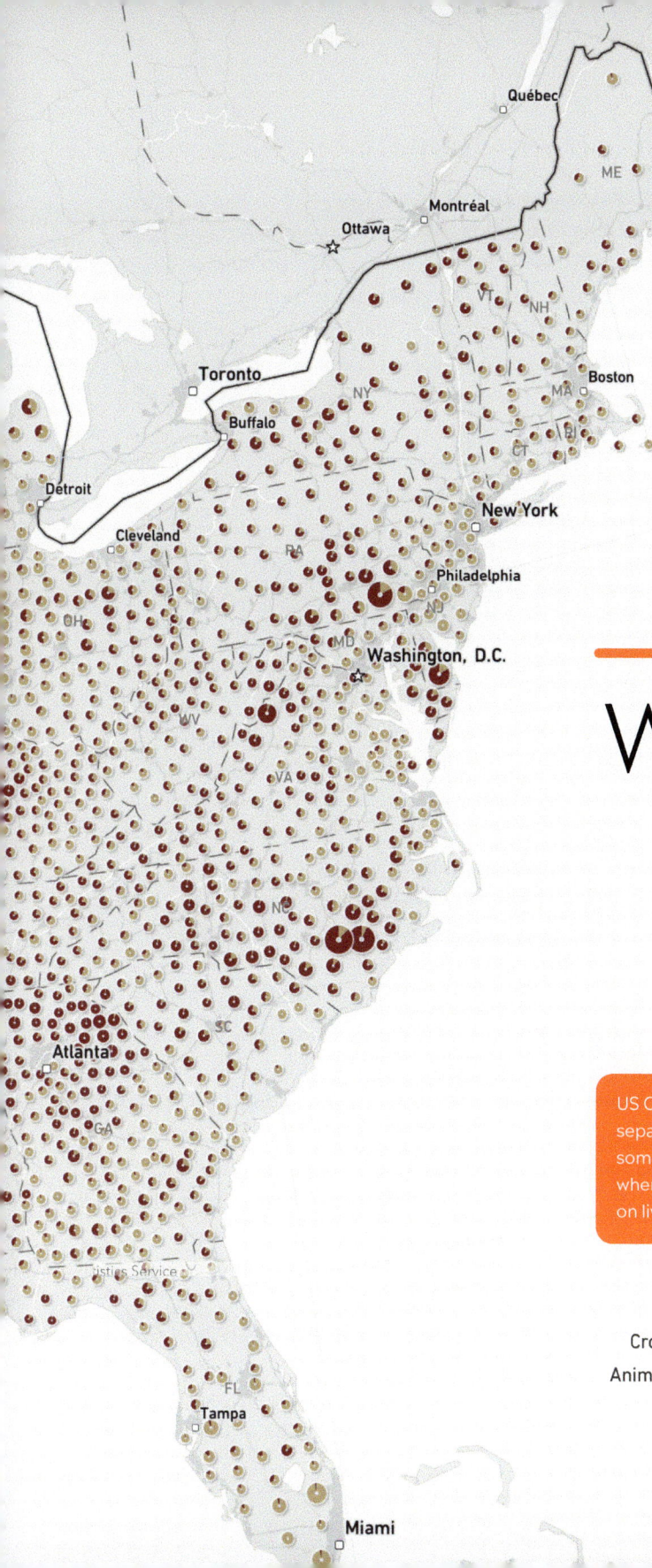

# 2

## Why Maps Matter

US Census of Agriculture data shows that, by separating crop sales from animal product sales, some regions (like the Midwest) specialize in crops, whereas others (like northern Texas) instead focus on livestock and other animal goods.

### Total agricultural sales, by category

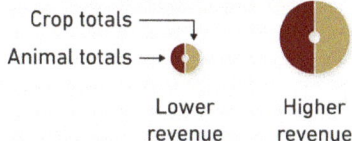

Crop totals →
Animal totals →

Lower revenue

Higher revenue

# Organizing information

All of us, as individuals and as groups, spend time and energy trying to organize and categorize information in one way or another—to discern relationships, prioritize tasks, update to-do lists, remember what to buy at the supermarket, and gain insights, both trivial and profound.

Designer, author, and TED Talks founder Richard Saul Wurman reminded us in his book *Information Anxiety* that there are essentially only five ways to organize information. He cleverly organized the ways information can be organized into this mnemonic:

|  |  |
|---|---|
| **LATCH** | **L**ocation |
| | **A**lphabet |
| | **T**ime |
| | **C**ategory |
| | **H**ierarchy |

We'll come back to location, but meanwhile let's start with alphabet. Alphabetizing can be a convenient tool for arranging and accessing things. But ordering items by their initial letters provides few insights and seldom indicates true relationships. Time helps us compare things and link events, but it's one-dimensional. Category can include elaborate and useful taxonomies, but categorization can be a subjective exercise. Although hierarchy helps us sort things, such as biggest to smallest and most expensive to cheapest, it feels simplistic and can also be subjective.

Location is unique in that nearly everything that's tangible has a location or geographic extent. As geographer Waldo Tobler stated in his First Law of Geography, though everything is related, the closer things are to each other, the more related they are. Location reveals commonalities and interrelationships. And location comes with its own visual language: maps. The language of maps is eloquent—even poetic—in translating, on paper and screens, the varied and tangled textures of our world into visual representations that reveal hidden patterns. Maps can also be the connecting tissue between location and time, using the vocabulary of cartography to represent temporal data with color fields, arrows, symbols, and animations.

# The power of maps in narratives

Maps add extra dimensions to multimedia stories. They pin narratives to place, situate a story within a larger context, and provide additional insights. In interactive form, they can enhance the viewing experience; whether interactive or static, they add to the variety and visual impact of stories.

Maps in narrative contexts help answer these and other important questions:

- Where am I?
- How do I get from here to there?
- What should I know about this place?
- Who lives here?
- What's near me?
- What used to be here?
- What will this place be like in the future?
- What's happening here?
- What might happen here?
- What's this place like compared with other places?
- Why should I care about this place?
- How can I help make this place better?

Maps play roles that other multimedia content just can't perform. They enrich the multimedia mix, and they provide an additional means of connecting to audiences. Maps can convey an immense amount of information with unparalleled efficiency, revealing patterns and interdependencies that photographs, video, and text are largely incapable of portraying.

Maps perform a variety of functions within a story. Their simplest task is to provide a single location, and thereby anchor a narrative to a place. Example: The simple globe locator at right pinpoints Palm Springs, California. In this case, the story within which it appears provides the context, negating the need for even a place-name label.

The next level is to depict a series or collection of locations. The map tour function within ArcGIS StoryMaps, like this panel from **Welcome to Palm Springs** ⊖⊃, locates a series of points of interest on a single map.

A globe locator of Palm Springs, California, in North America (*top*) and a map tour of the city (*left*).

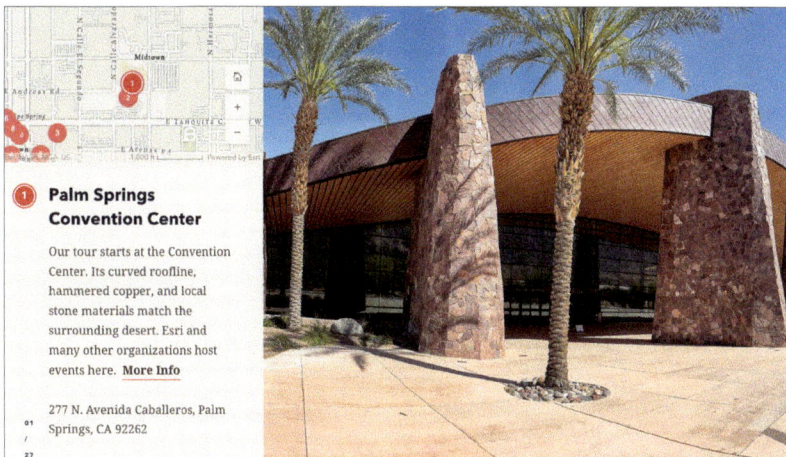

**1** **Palm Springs Convention Center**

Our tour starts at the Convention Center. Its curved roofline, hammered copper, and local stone materials match the surrounding desert. Esri and many other organizations host events here. **More Info**

277 N. Avenida Caballeros, Palm Springs, CA 92262

Understanding the distance and orientation of one point to others in the narrative provides additional richness and context. It's more than just dots on a map—it's a cartographic tour guide.

Thematic maps—singly, and especially in series—can parse a complex tapestry of categories and relationships. Example: **The Lands We Share** ⊖ story (*below*) presents and interprets maps of protected lands and waters in the United States, depicting levels of protection and patterns of land management.

Maps in series or presented as an animation can vividly depict change over time. **An Introduction to Sea Ice** ⊖ (*bottom*) uses both devices to show seasonal change in ice extent and long-term reduction in sea ice caused by climate change.

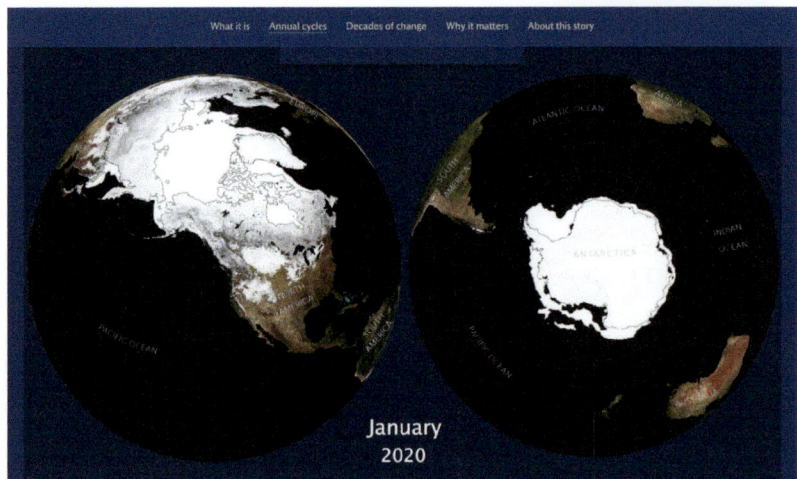

*The Lands We Share.*

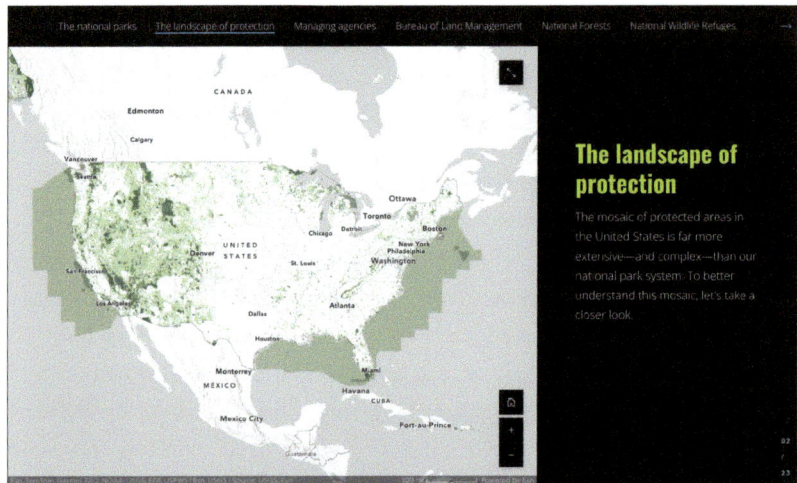

*An Introduction to Sea Ice.*

Maps provide context within stories. In many cases, maps form the heart of a multimedia story. Many story ideas *begin* with a map, or a series of maps and layers, often as a result of a GIS analysis, that reveals patterns, depicts trends, or otherwise demands interpretation. Just as maps can provide context to stories, stories can provide context for maps.

We'll further explore how maps *perform* within stories in chapter 6.

# Three storytelling biases: Word, picture, map

All of us, I suspect, have biases, whether conscious or unconscious. We drag them along with us, like business travelers wearily rolling suitcases through airports. Biases are often a consequence of the way we've trained our minds to work—and that, in turn, often has to do with our social and cultural backgrounds, our academic training, and our professional lives.

Some of us consider ourselves word people; some are picture people. Others of us think spatially. These predilections affect the way we learn. They also influence how we communicate and how we approach storytelling. A comedian, a tax attorney, and a sculptor will almost inevitably have strikingly different storytelling styles.

Multimedia narratives, because they combine text, visuals, and maps, pose special challenges—and opportunities—for us as storytellers. Our deeply ingrained habits can severely limit how we approach telling stories. A word person might consider visuals as an afterthought; a cartographer might give text and pictures short shrift. Transcending our natural biases and embracing diverse forms of storytelling can be liberating, mind-expanding—and fun. So let's examine three biases that, if consciously and effectively transcended, can ensure captivating and beautiful multimedia storytelling.

## The *word* bias

Traditional newspaper reporters (admittedly an endangered species these days) conduct research and do interviews with an end in mind: a text piece of a few hundred words that they hope will make the front page. All the reporter's energy is poured into gathering the five *W*'s and one *H*—who, what, when, where, why, and how—that will make for an impactful, informative story. In traditional, print-media storytelling, a photographer might occasionally accompany a reporter in the field, or a photo editor might

acquire an image after the fact and as deadlines loom. But words over-whelmingly bear the storytelling burden.

Regardless of the circumstances, a writer will most likely conceive, research, report, and deliver a story with an unconscious bias in favor of things that can be easily described in text. Skilled writers can, with words alone, paint a detailed picture of a scene or environment. They can describe individuals and interactions; they can even move us to tears. But all too often, reporters minimize—or perhaps omit altogether—elements of the story that don't lend themselves to text descriptions.

As a mental exercise, let's consider how effective words, images, and maps are at tackling the five *W*'s and one *H* of reporting. I'll use dots to roughly measure effectiveness: more dots indicate more effectiveness. This is a somewhat subjective exercise; you'll probably assign things differently than I. My feeling, though, is that text is particularly good at *what, why,* and *how. When* and *where* can be simply described with a place-name and a date, but that doesn't provide much insight. Sure, a good writer can, with effort, vividly describe a personality, but text's great strength is in expo-sition, which is defined by the *Oxford English Dictionary* (*OED Online*) as "a comprehensive description and explanation of an idea or theory."

| | Who | What | When | Where | Why | How |
|---|---|---|---|---|---|---|
| Word | ●● | ●●● | ● | ● | ●●● | ●●● |

An example of a word-driven story: **University Libraries as Providers of GIS Services: A Guide** ⊝, by David Cowen, distinguished professor emer-itus of geography at the University of South Carolina, is a digital book in the form of a collection of ArcGIS StoryMaps stories. Its case studies and best practices are aimed at a professional audience of librarians, scholars, and administrators. Thus, the length and larger proportion of text seems appropriate.

But the collection also includes visuals. It's advisable to break up text now and then with section headers, callout quotations, and other elements, such as tables and infographics. Readers of even the most technical story will be grateful for occasional images or other visual breaks in a long nar-rative. That said, visuals shouldn't simply be cosmetic; they should support and complement the narrative.

## University libraries as providers of GIS services: A guide

David Cowen, Professor Emeritus, University of South Carolina

This ArcGIS StoryMaps collection is intended for anyone charged with providing GIS technical support and instruction for faculty, staff, and students at colleges and universities. It is a free online document that allows readers to easily navigate through chapters and learn from experts who have successfully implemented GIS support programs at their institutions.

Introduction
and contents

A tour of GIS programs
Chapter 1

A brief history
Chapter 2

Why are libraries ideal?
Chapter 3

Recipes for success
Chapter 4

Trends and directions
Chapter 5

Challenges
Chapter 6

Story collection: *University Libraries as Providers of GIS Services.*

## What is special about data libraries?

Starting with the GIS Literacy Project in 1992, academic libraries have been viewed as a potential setting for cross-campus GIS support. Initially, census files on CD-ROM provided a basis for free nationwide GIS data. As a stand-alone system, this data was well suited to a library's government documents department. As we moved from fixed media to web-based data portals, we found new ways to discover data and create customized GIS applications. The ability to scan existing paper maps in a map library offered a new opportunity to convert map collections to digital data repositories. When Ann Holstein conducted a major survey in 2015, she optimistically concluded:

> Receiving more funding will mean more staff, better-trained staff, a more in-depth collection, better hardware and software, and the ability to offer multiple types of GIS services.
>
> *Holstein*

van Brakel and Pienaar (1997) recognized the important role that libraries could play in the future of GIS:

> Libraries are the primary organizations for processing, organizing, and disseminating information and as dynamic organizations they must evolve through the adoption of new services such as GIS that will enable them to provide their users with innovative services to meet ever-increasing needs.
>
> *Van Brakel and Pienaar*

J. Benner and E. Slayton (2020) also provide a rationale for an expanding role for libraries. They suggest that "At their core, all libraries are cultural heritage institutions that support education, research, and civic life within a community." Libraries act as a stronghold for education on spatial literacy and critical engagement with geographic concepts.

The literature on the development of academic GIS support programs highlights the advantages of providing GIS these services in libraries. These advantages can be divided into three categories:

### 1. Location/mission

- Libraries as a place for accessing knowledge—including maps and spatial data
- A common space on campuses to foster this knowledge, and the central position of academic libraries as a campus unit that serves everyone,
- A central location on the campus and succeeds with collaboration—in the library and across campus
- An engaged, neutral service provider
- A communal resource for people on campus needing GIS and data services
- A unique position as a place of learning that exists outside individual departments and engages in instruction across the university
- A learning space for building skills in research and critical thinking
- Supports intellectual freedom for individuals to have free and open access to resources and services for undertaking a new information pursuit (Boxall)
- Consults with and supports individuals in their existing and emerging projects
- Connects people in meaningful ways using their thorough understanding of users' skills and needs gathered through intensive research consultations
- Works continually to improve service models and devises strategies that integrate evolving technology and accommodate growing demand for geospatial resources

### 2. Resources

- Develops collections of GIS-related materials and provides access to geospatial data and software tools
- Provides effective workspaces, software, hardware, data sets, technological expertise, and training
- Assembles GIS services, including training, in a common space on campus
- Provides an ideal place for those whose disciplines do not have a large focus on data or GIS

### 3. Supports spatial literacy

- Serves numerous roles in providing GIS services through analysis and response to a nontraditional set of user needs
- Uniquely situated to act as a springboard for teaching spatial literacy and critical engagement with geographic concepts
- Connection points to help community members and students learn about each other, spatial concepts, and geospatial tools
- Supports instruction on GIS and spatial literacy
- Draws on the experience of GIS and data services librarians
- Provides outreach to a variety of communities about geographic concepts and the role of geography in society (Benner and Slayton)

## The *image* bias

Photographers and videographers, in contrast to writers, will typically take different approaches to the same story. Depending on deadlines, budgets, and the constraints of the publishing format, they may be seeking to make a single image that symbolizes or encapsulates a story. Or they may have the luxury of publishing a series of images or videos that together form a visual narrative.

There are many approaches to image-making. One is the traditional photojournalist's and street photographer's mantra that I learned at National Geographic: "F/8 and be there." In other words, fret less about the technical aspects of photography (just put it on a standard F-stop), concentrate on being in the right place at the right time, and have an alert and perceptive frame of mind.

Another photographers' dictum: "The decisive moment." Capture a fleeting instant where elements come together in a perfect composition. The French photographer Henri Cartier-Bresson famously captured decisive moments, such as a man frozen in midair as he leapt across a puddle, or a cyclist as he raced down a cobbled street.

A captured instant rarely tells a complete story. But a series of images can tell a story, and can enrich a multimedia story in at least three ways:

- It can provide a vivid sense of place. Text, too, can evoke a sense of place, but images do so with greater efficiency. A glance at an image of a desert or rain forest thrusts us instantly into an exotic environment.
- It can introduce us to people. Human brains are wired to respond powerfully to human faces. Seeing a story's hero depicted in a photograph makes us more likely to empathize with that person.
- It can elicit emotional responses. An image of a dead elephant with its tusks sawed off by poachers can affect us much more powerfully than a written description of a threatened species or illegal wildlife trade.

Most of us snap pictures with our smartphones when we're on vacation. But the resulting collection of images rarely tells a story beyond a mere visual checklist of selfies, scenic views, and family portraits. The challenge for a visual storyteller is to go beyond the superficial pretty pictures to attempt to capture the spirit and essence of a place. What tells you more about Paris: a snapshot of the kids in front of the Eiffel Tower or vignettes at sidewalk cafes and an atmospheric shot of the Seine coursing through the City of Light? Veteran photographer Jim Richardson, who has

Images of Paris: tourist snapshot (*top*) versus storytelling (*bottom*).

photographed dozens of stories for *National Geographic Magazine,* describes how, early in his career, he came to realize that his job wasn't to photograph things; it was to photograph *ideas.* He would work with a picture editor to thoroughly research a story concept, and then distill the key elements of the story into a series of topics. The challenge was to turn those topics into original, memorable images.

What are the storytelling strengths of images? In a single image, a skilled portraitist can convey a vivid sense of *who*; and, of course, a street or landscape photo can transport us to *where*. But photographs can often fall short when it comes to the important elements of *what*, *why*, and *how*.

|  | Who | What | When | Where | Why | How |
|---|---|---|---|---|---|---|
| Image | ●●● | ● | ●● | ●●● | ● | ● |

For examples of effective image-based stories, you need go no further than the International League of Conservation Photographers, a group I've long admired. Their members have created many multimedia stories with compelling images. An example is **Choking on Convenience** ⌒⊃, which vividly depicts the devastating impact of plastics on the natural world.

*Choking on Convenience* from the International League of Conservation Photographers.

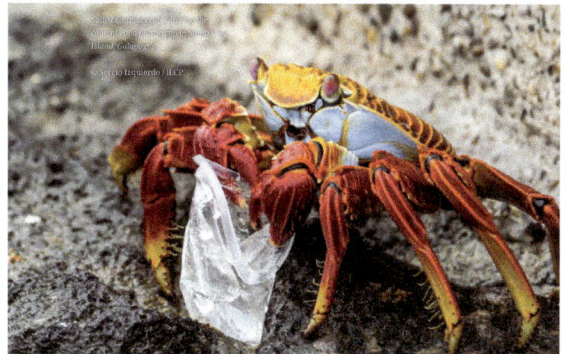

**Choking on Convenience**

iLCP Photographers Unveiling the Silent Threat of Plastic Pollution

By Meg Severide, Earth Day 2024

Manila

PHILIPPINES

**Plastic Coral**

*Maricaban Island, Philippines*

Off of Maricaban Island in Batangas, along the Verde Island Passage, which is known to host some of the most biodiverse marine ecosystems on the planet, a seemingly harmless algae-covered plastic bottle is wedged between coral and feather stars. Plastic pollution can persist for years

*A young brown fur seal with a fishing net entangled around its neck and slowly tightening. Cape Cross colony, Namibia*

© Emanuele Biggi / iLCP

The story includes an interactive map tour that demonstrates the issue's worldwide scope.

## The *map* bias

Cartographers think about how they can visualize spatial data in ways that reveal patterns, show interrelationships, and tell stories. They're aware of the simple role that locator maps can play to place a story in a geographic context. But mappers really love to tease insights out of the complex and seemingly chaotic elements and phenomena that make up a place or region.

Maps are models of the world (or a portion of it); they simplify by representing a few key items and eliminating nearly everything else. Their language is in symbolization, using colors, lines, tints, vignettes, and icons. For cartographers, the tricky part of storytelling is creating a narrative. Traditional printed maps can create subtle narratives through visual hierarchies, manipulating colors and typography to guide readers into and through a map. Digital maps published in ArcGIS StoryMaps or other electronic media can make narratives more literal—and maps easier to interpret—by revealing data layers one by one, as in the sequence on the facing page.

A map's greatest storytelling strength is in depicting *where*. Maps, especially in series, can show change over time, so I think it's fair to consider *when* as another core strength. I could easily have added a third dot to *why* and *how*, because maps can be so effective at depicting the relationship of one phenomenon to others.

| | Who | What | When | Where | Why | How |
|---|---|---|---|---|---|---|
| Map | ● | ●● | ●●● | ●●● | ●● | ●● |

An exemplary map-based story is **American Agriculture by the Numbers** ⊖, created by Cooper Thomas and other members of my team. As the story's introduction explains, "Every five years, the US Department of Agriculture invites farmers and ranchers across the United States to complete the Census of Agriculture. The results—an official set of uniform agriculture data for every state and county—offer insights into the landscape of farming and ranching nationwide." Maps predominate in this story, but they're supported by a mix of text, images, and infographics. Words and images alone cannot reveal patterns that are instantly discernable within the story's nine maps.

**The next generation**

This is no coincidence. To address population aging among agricultural producers, many Midwestern states offer incentives to young farmers contemplating entry into the industry. As a result, young farmers make up a greater proportion of the agricultural workforce in these regions.

This map, which shows the total number of producers under age 35 (circle size) as well as their proportion relative to all producers (circle color), size reveals some curious outliers. For example, the three counties with the most young producers — Lancaster County, Pennsylvania; LaGrange County, Indiana; and Holmes County, Ohio, respectively — are also home to the three largest Amish populations in the country.

Total producers under age 35

Proportion of producers under age 35
Lower proportion — Greater proportion

**Farms and ranches are distributed throughout the country**

This map, which shows the total number of agricultural operations by U.S. county, captures the overall geography of agricultural operations nationwide.

The distribution of farms and ranches roughly hews to the physical geography of the country, with operations scarce in mountains, deserts, and other inclement environments.

The densest concentration of farms and ranches extends down the spine of the American heartland — from Minnesota in the north to Texas in the south — but agricultural operations are also clustered in parts of the Far West, across Florida, and astride the Appalachian Mountains.

Total agricultural operations, by county
Fewer operations — More operations

**Acreage per operation**

Smaller — Larger

**Average producer age**

Lower revenue — Higher revenue

**Agricultural sales**

Younger — Older

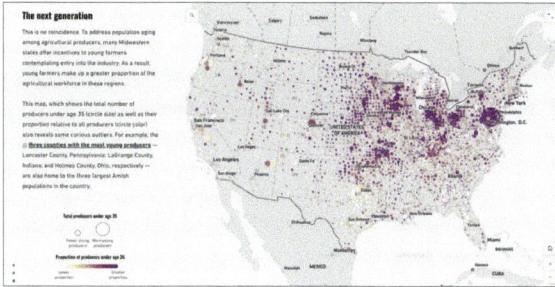

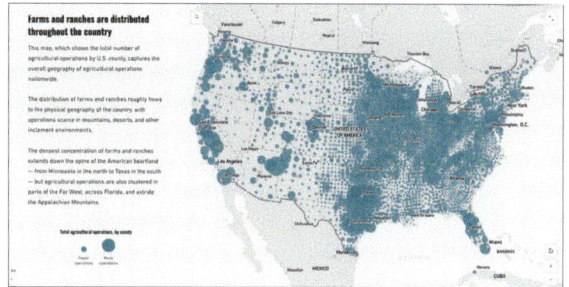

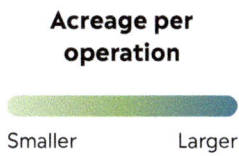

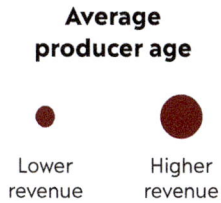

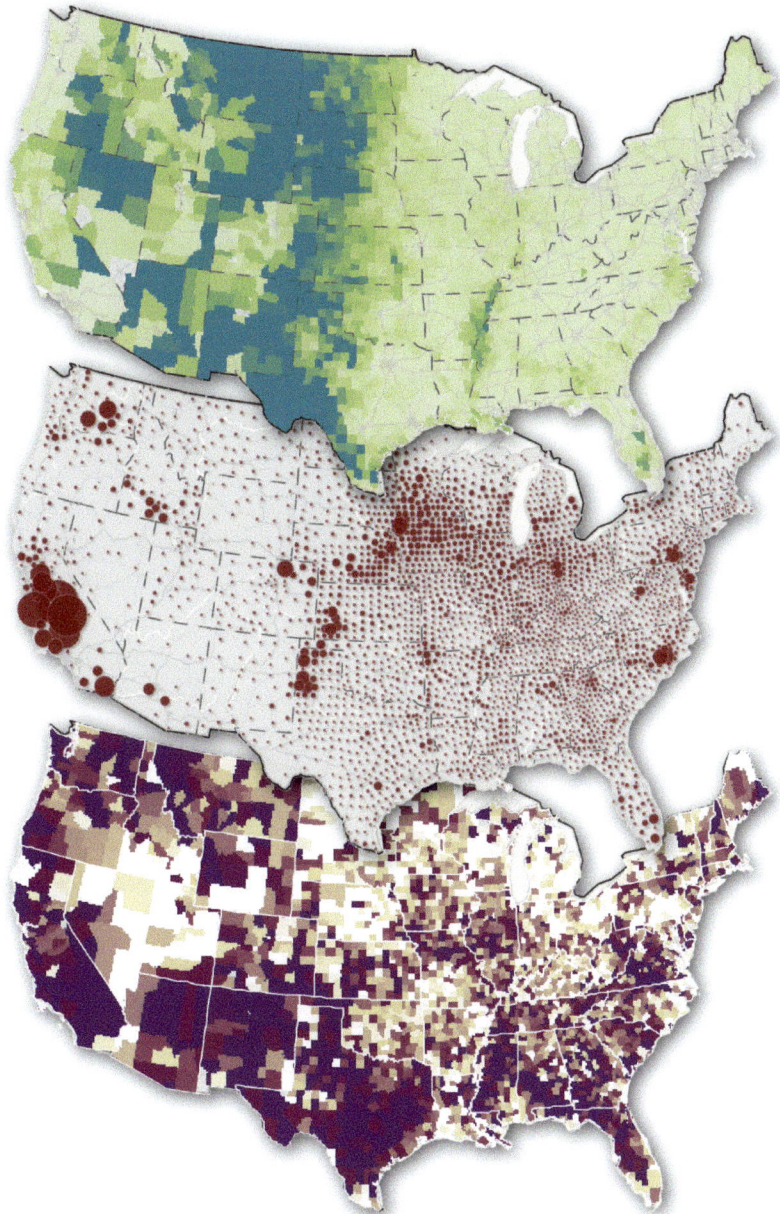

Mapping the state of agriculture in the US.

# Combining words, images, and maps

Words, images, and maps are all worthwhile storytelling devices. But they all have limitations, as well. The magic of multimedia stories is that they benefit from all three media, all three modes of thought. The challenge is that few people have the training—or have opened the very different neural pathways—that enable and encourage creative thinking along all three lines: textual, visual, spatial.

So let's evaluate the relative strengths, together, of word, image, and map.

You might quibble with the details, but words, images, and maps, in aggregate, cover all the storytelling bases. I compiled these little charts separately as I drafted this text. But I was pleasantly surprised to realize that, when word, image, and map are combined, nearly all the categories add up to the same number. Only *where* exceeds the rest, and that's entirely consistent with our passion for ArcGIS StoryMaps—my team and I have a quite conscious bias toward maps.

| | Who | What | When | Where | Why | How |
|---|---|---|---|---|---|---|
| Word | ●● | ●●● | ● | ● | ●●● | ●●● |
| Image | ●●● | ● | ●● | ●●● | ● | ● |
| Map | ● | ●● | ●●● | ●●● | ●● | ●● |
| Total | 6 | 6 | 6 | 7 | 6 | 6 |

Innumerable multimedia stories combine words, images, and maps to tell memorable and inspiring stories. A particularly effective one is **A River Interrupted** ⇔, winner of Esri's 2023 ArcGIS StoryMaps competition. It opens with drone footage of one of the many dams on the Charles River in Massachusetts. The video and title complement each other, pulling readers into the story.

Within the narrative, introductory text lays the groundwork for the story. Photos and videos within immersive sections include archival images and maps, providing historical context. Choreographed map series locate 19 dams along the river and show how they have blocked the migration of shad, alewife herring, blueback herring, rainbow smelt, and Atlantic salmon. Infographics illustrate other environmental impacts of dams. Callout quotes enliven the text. Swipe functions illustrate the before-and-after benefits of river restoration. An option to "unmute background audio" activates environmental sounds of flowing water. And a final section invites readers to "join the movement for dam removal." You may suspect that the result is a confusing mix of warring elements, but the effect is instead visually harmonious and easy to navigate—and it drives home a strong and convincing case for restoring the natural flow of the river.

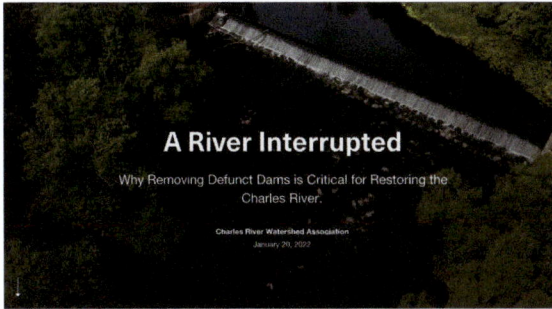

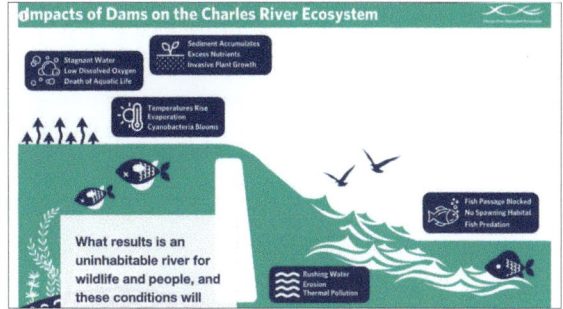

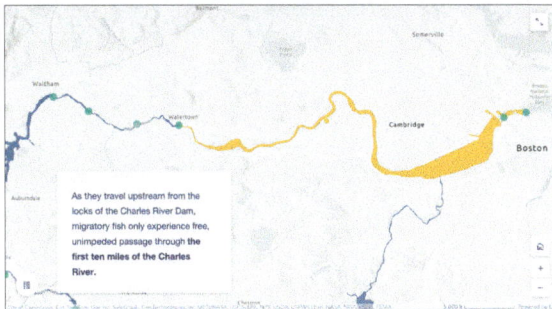

*A River Interrupted,* exploring the impact of dams on the Charles River, Massachusetts.

# What about sound?

We've discussed words, images, and maps at length. But *A River Interrupted* reminds us that audio is another important element of the multimedia mix. Audio tends to be undervalued as a medium, but adding environmental sounds to stories, especially when image and audio are experienced simultaneously, can viscerally evoke an environment.

Our team's first real effort to incorporate sound was a collaboration with the Acoustic Atlas on **Sounds of the Wild West** ⊖. The story featured landscape photos of the western United States with appropriate aural accompaniments: howling wolves, warbling meadowlarks, bugling elk, and bubbling Yellowstone mud pots. Modern web browsers require readers to manually opt in to sound, but once readers opt in, audio within the main story, sidecar, and slideshow immersive elements plays automatically as readers (listeners) scroll.

Authors creating narratives with ArcGIS StoryMaps can upload audio directly within the story builder; an additional option is to embed audio players into the story. Clicking a player is less elegant than having sound occur automatically as readers scroll. But clicking a player and hearing a grizzly roar, as occurs within the *Sounds of the Wild West* story, is a startling—and effective—experience.

# World's Longest Mule Deer Migration

## Red Desert to Hoback

**Authors:** InfoGraphics Lab, Department of Geography, University of Oregon in partnership with the Wyoming Migration Initiative

**Medium:** ArcGIS StoryMaps

**Story behind the story:** In 2016, a mule deer doe, named Deer 255 by researchers and outfitted with a tracking device, migrated 242 miles from eastern Idaho to southwestern Wyoming, crossing corners of Yellowstone and Grand Teton National Parks, Bridger-Teton National Forest, and a variety of BLM and private lands.

She passed through several choke points, including a tract of land near Pinedale, Wyoming, that was slated for development. Partly based on tracking data from multiple animals, the Conservation Fund purchased the tract and donated it to the Wyoming Game and Fish Department.

**Why it's special:** Deer 255's migration forms the centerpiece of a multimedia story highlighting multiple migration routes and calling attention to increasing threats that wildlife face from roadways and fence lines as development spreads into formerly open rangelands.

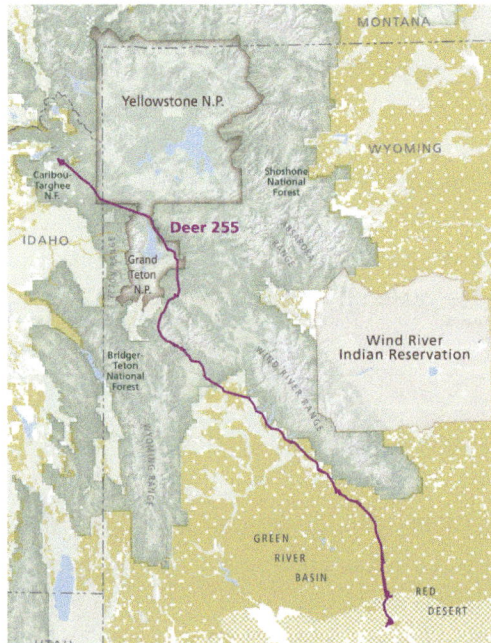

*Above*: A migrating mule deer stops for a snack.

*Left*: Deer 255's odyssey took her across lands managed by a variety of agencies.

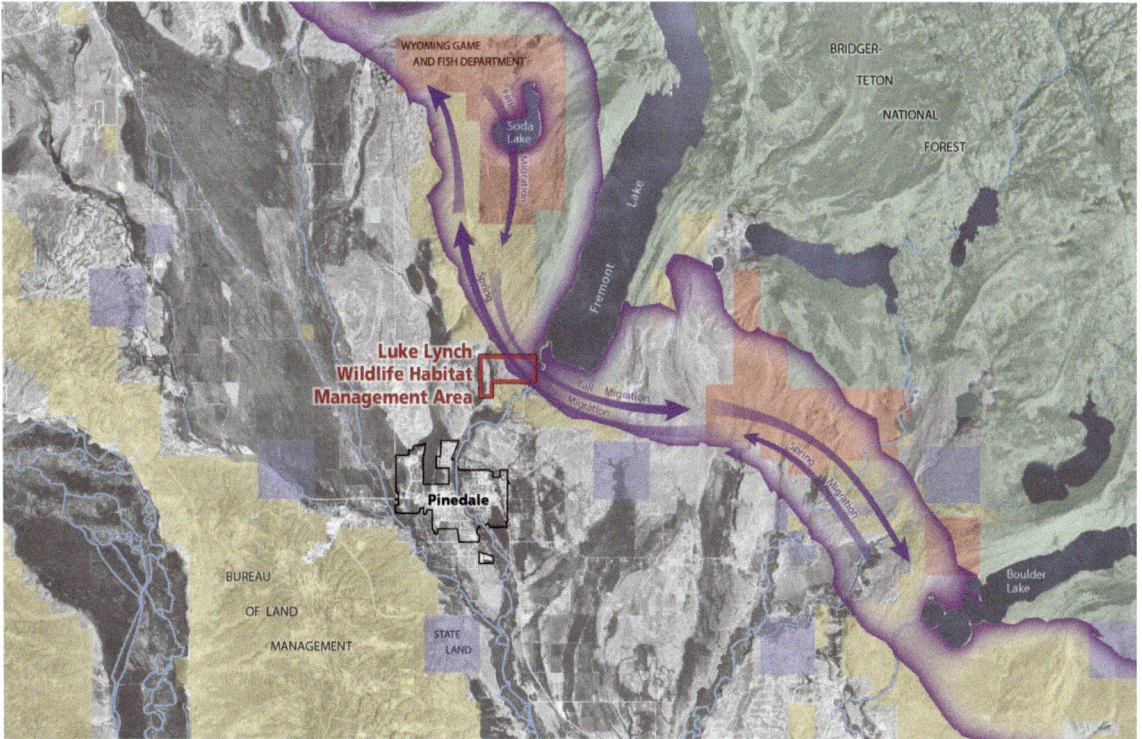

Migration maps (*top*) helped identify a choke point that was slated for development. After the parcel was acquired, volunteers helped remove a fence (*left*). An increasing number of wildlife overpasses and underpasses are helping reduce animal fatalities at critical road crossings (*above*).

# Malaria on the Frontlines

## An interactive tour of the communities fighting to end malaria for good

**Author:** The United Nations Foundation's United to Beat Malaria campaign

**Medium:** ArcGIS StoryMaps

**Story behind the story:** "The mosquito is the deadliest animal in the world," aptly declares the UN Foundation in this ArcGIS StoryMaps story, adding that malaria, a mosquito-borne disease, killed more than 400,000 people in 2019.

**Why it's special:** The story uses a map tour format to profile individual activists fighting the disease on both sides of the Atlantic. Elsewhere in the story are case studies from Ecuador, Haiti, Uganda, and Benin. Photos, video, and audio create an immersive experience, and a consistent focus on individual people makes the story especially engaging.

The narrative ends with clear calls to action and invites readers to participate through social media.

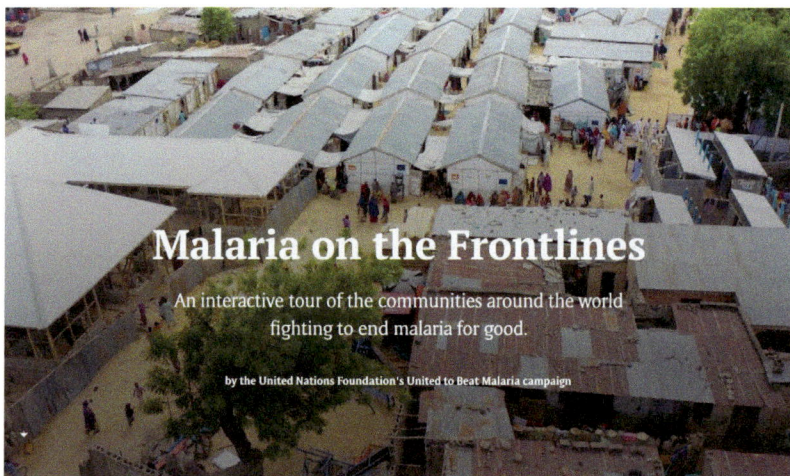

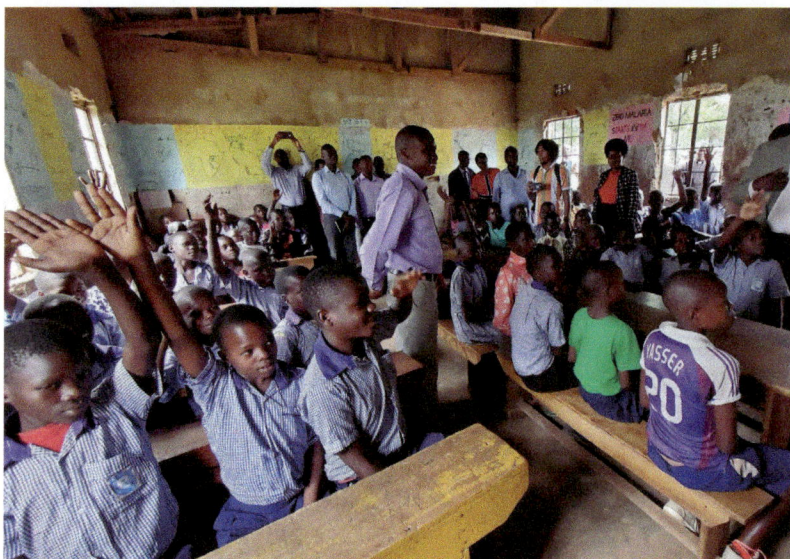

Malaria on the Frontlines

An interactive tour of the communities around the world fighting to end malaria for good.

by the United Nations Foundation's United to Beat Malaria campaign

*Top*: Cover art and title. *Above*: Young students learn about malaria in a western Ugandan classroom. Kids recite in unison and applaud in an accompanying audio clip.

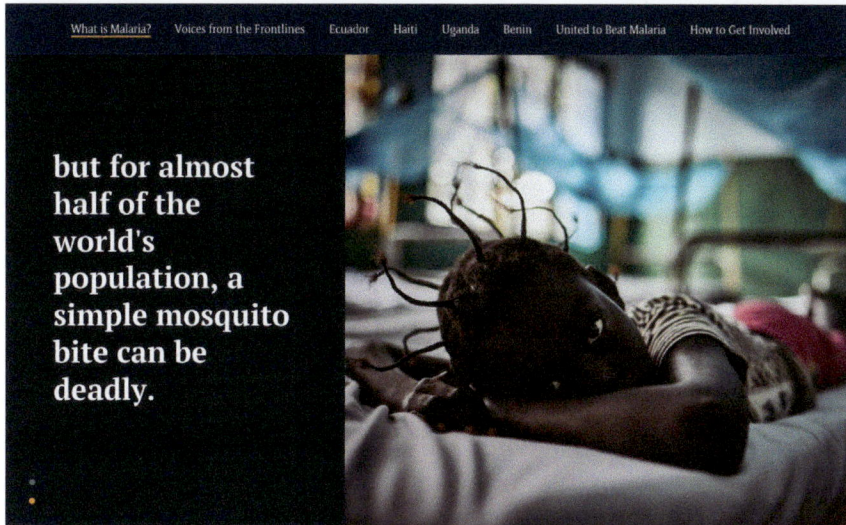

*Above*: *Voices from the Front Lines* tours Latin America and Africa, profiling health professionals combatting malaria.

*Left*: An image of a young malaria victim adds emotional impact to the story.

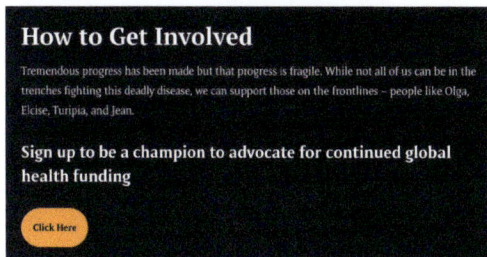

Buttons encourage advocacy and financial support.

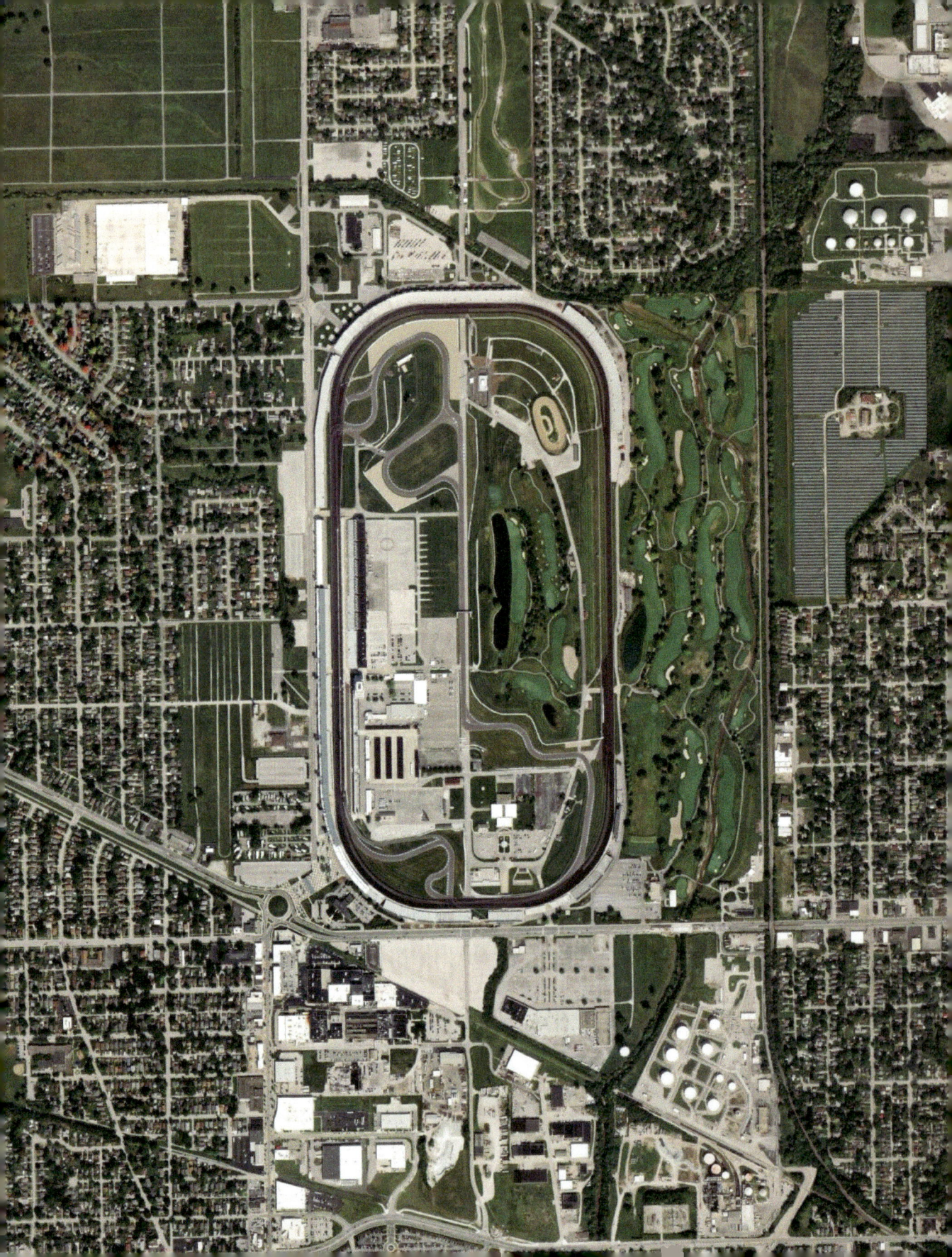

# 3

## Maps and Minds

High-resolution imagery shows the two-and-a-half mile oval of the Indianapolis Motor Speedway nestled within the city's west side street grid.

# Maps and memory

I grew up in the middle: a middle child in a middle-class family in a comfortable midsize town more or less in the middle of the country. My stomping grounds covered the north side of Indianapolis, which back then was a sleepy, midsize, Midwestern metropolis.

And yet this rather undistinguished landscape imprinted itself so profoundly on my young mind that for years after I left the city, most of my dreams were staged there. Now and then I still have Indianapolis dreams. Perhaps that's unusual, but I suspect that, for many of us, childhood memories are intimately intertwined with very specific locations that remain vivid in our minds for the remainder of our lives. Memories are bound to locations, and locations, conversely, evoke memories.

The map below shows the geography of my youth, annotated with places that I can still picture vividly and that recall indelible childhood memories,

The near north side of Indianapolis.

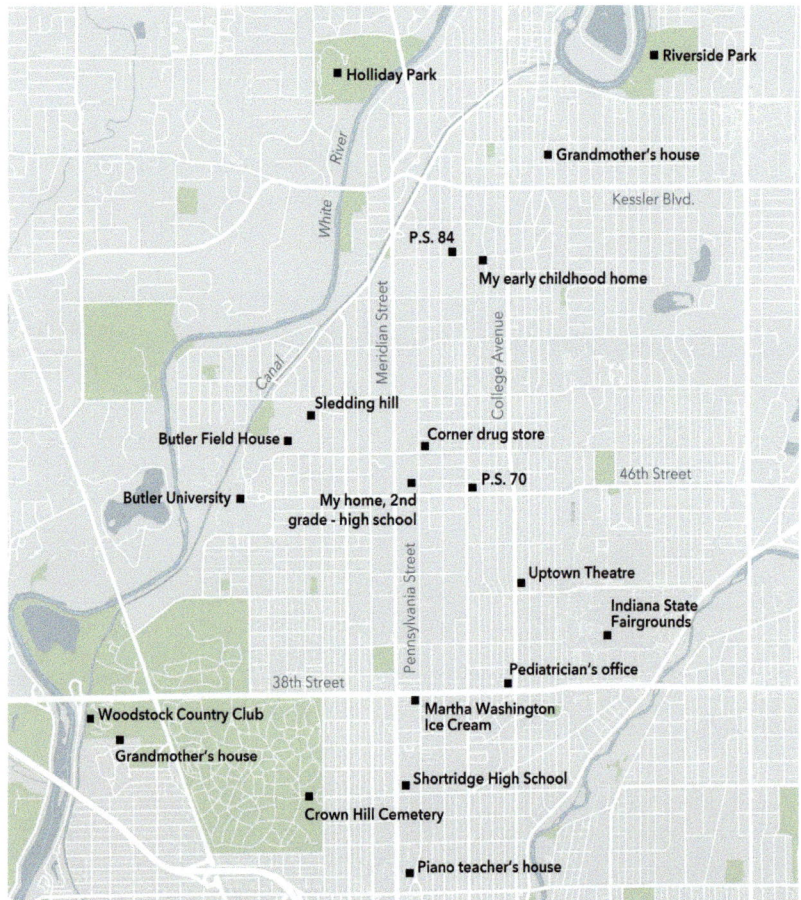

The same geography, annotated with locations of childhood memories.

from the thrill of a sled ride down one of the city's few hills—at the edge of the Butler University campus and now obliterated by student apartments—to anxiety-provoking visits to the pediatrician's office at 38th Street and College Avenue. To nearly all who read this book, these places are just labels on a map. To me, each location carries deep significance and conjures vivid images. I can still savor the hot fudge sundaes at Martha Washington's on 38th Street, even though the ice cream parlor disappeared decades ago. I can envision every block of my commute by city bus from my home on Pennsylvania Street to Shortridge High School, at 34th and Meridian Streets. I can hear the cicadas buzzing in the backyard of my grandmother's house on Golden Hill Drive.

Similarly, when I recall individual childhood memories, they seem to have mental maps attached to them. When I conjure up recollections of attending Saturday matinee showings of Three Stooges movies at the Uptown Theater, they're semiconsciously pinned to a location relative to other childhood locations, other memories, on a map floating dimly in my head. Do all of us form mental maps when we think of the past? Or is it the province of those of us who are more spatially oriented?

Obviously, our ability to recall places—and to maintain mental maps—is somehow closely linked to our ability to remember events and episodes in our lives. But as I've pondered those links, it's led to other questions, among them:

- How do we remember and recall locations and routes, whether it's the few steps across the kitchen to the drawer where the measuring cups reside or the fastest route to the local supermarket or the relative locations of city landmarks?
- How do we (and animals of all sorts) navigate through space?
- Do we have maps in our brains? How can a tangle of neurons build and maintain a mental map?
- How do we go about making sense of the world?

So I did a little reading, starting with a *Scientific American* article titled "The Brain's GPS Tells You Where You Are and Where You've Come From."

## Maps and the brain

It turns out that much of our spatial thinking machinery is housed in the inner, more "primitive" region of our brains. That makes sense, since our evolutionary ancestors have had to navigate across landscapes—and seascapes—for millions of years. It should also come as no surprise that the structure and mechanics of spatial memory and cognition are complex.

They involve an intricate interplay among specialized cells in various parts of these inner regions, with abundant links to our more recently evolved outer cortex.

Researchers have discovered that we possess something very much like a physical map—or series of maps—deep within our brains. According to May-Britt Moser and Edvard I. Moser, Nobel Laureates and professors in psychology and neuroscience at the Norwegian University of Science and Technology in Trondheim, nerve cells in the brain fire in a way that indicates the layout of the environment and how an organism is positioned in it. Our brains contain arrays of grid cells that are stacked in layers corresponding to various scales in the physical world. Arrays near the top of the stack are spaced more closely together and aid navigation over shorter distances. As we move across a room or a landscape, these grids apparently light up in patterns that approximate our real-world movements.

Schematic representation of grid cell arrays within our brains.

The arrays of grid cells reside in a part of our brain called the entorhinal cortex. These arrays communicate with clusters of specialized cells in the hippocampus, which also functions as a kind of memory processor. This portion of the brain integrates and consolidates information from our short-term recollections and files it away as longer-term memories. Among these specialized cells are place cells that store information about specific locations, while others perceive additional components of our movement through space. Further investigation on lab animals has revealed the presence of head direction cells that fire according to our orientation. Border cells fire as they near boundaries—walls, fences, and such. Various speed cells fire according to how fast the subject animal is moving. Somehow they integrate the inputs from all these specialized cells and arrays and manage to navigate the world, and their homes, without getting lost—usually.

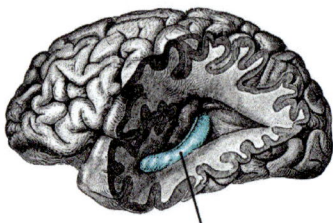

The hippocampus, nestled deep within our brains, is key to our spatial cognition.

Another of the hippocampus's many roles is to assist in the processing of emotions. It's perhaps no wonder, then, that my spatial memories of Indianapolis have such a strong emotional component.

London's cab drivers—famous for "the Knowledge" nickname for their command of the city's vast maze of streets—apparently have larger hippocampi than the average Londoner. And experienced cabbies tend to have larger hippocampi than their rookie colleagues.

## Maps and learning

Maps, and learning about mapping technology, can literally change young minds. Bob Kolvoord, interim provost and professor at the College of Integrated Science and Engineering at James Madison University, is founder of the Geospatial Semester, a program offered in Virginia high schools in which students learn to use the tools of GIS as they explore local issues. More than 7,000 students have been exposed to geospatial technology over the program's two decades of existence. When educators asked Kolvoord about the impact of the program, he realized that he had little or no data to demonstrate its effectiveness. Thus he began to look for ways to quantify the benefits of the program in collaboration with David Uttal, professor of psychology at Northwestern University.

Some initial studies indicated that students who had taken the course had an enhanced ability to use spatial language and showed improvements in broader verbal skills, such as defining problems and arguing from evidence. In addition, their work with GIS tools and concepts eliminated the small but significant differences in spatial skills between boys and girls, with girls catching up to their male peers.

In pursuing additional data, Kolvoord teamed up with Adam Green, a professor at Georgetown University and director of its Lab for Relational Cognition. Green's interest is in human creative intelligence; he had been doing research using functional magnetic resonance imaging (fMRI) to map brain activity. The Research Team (2022) recruited a group of students who were participating in the Geospatial Semester and a second group of kids from similar backgrounds who were not enrolled in the class. The groups underwent brain scans at the beginning and end of the school year, looking to see which areas of their brains were activated as they performed verbal reasoning problems. (A simplistic example: If X is better than Y, and Y is better than Z, is X better than Z?) Although the groups had nearly identical results at the beginning of the year, differences were apparent at year's end.

Green explained that they were looking for changes in brain regions that had long been associated with spatial cognition, especially a part of the brain called the posterior parietal cortex. They found that "the more students increased in activity in posterior parietal cortex, the better they

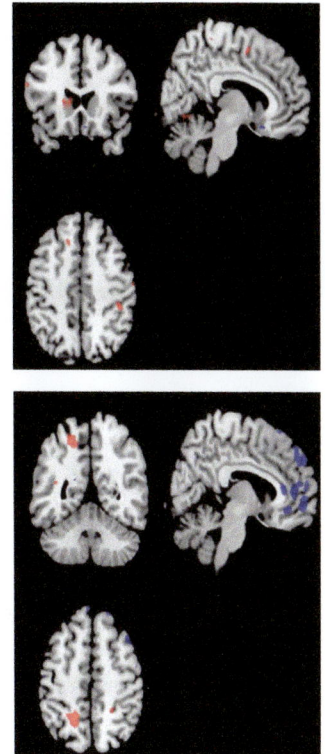

fMRI scans showing differences in brain activity while performing tests before taking the Geospatial Semester (*top*) and after (*bottom*).

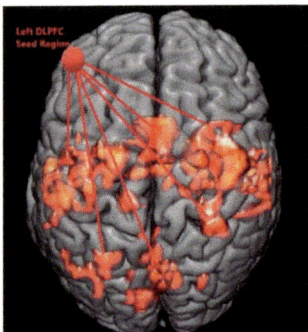

Increased connectivity among brain regions of Geospatial Semester students.

got not only at spatial cognition but other forms of reasoning as well," Green said.

"The students who did the Geospatial Semester used the spatial thinking parts of their brain preferentially when they had nonspatial problems," Kolvoord explained. "We're really intrigued by that, because it shows evidence that there's more to the use of GIS in a school setting than simply learning the material better."

Green agrees, explaining that the brain activity of the Geospatial Semester students wasn't just different, but a close look at the complex patterns of brain activity within spatial brain regions showed that students who took the Geospatial Semester started "spatializing" nonspatial forms of reasoning—that is, they started thinking about words the way that most brains think about objects. "The scans showed increased activity in their parietal lobes and decreased activity in their frontal lobes," Green said. "They're recruiting different parts of their brain to do these verbal reasoning tasks."

The scans not only showed activity in certain areas, but a greater connectivity between different parts of their brains. In other words, different parts of their brains worked together to a greater extent.

Green drew three conclusions from their research. First, "spatial training, spatial education, matters for a lot more than just looking at maps." Second, neuroscience could help educators plan and assess school curriculums by using fMRI to see how young minds are changed.

Third, "schools and school systems should value spatial education."

"The point of the Geospatial Semester is not for it to be a GIS-1 class," Kolvoord said. "It's to help students appreciate and practice spatial thinking and to use it in spatial problem solving."

## Maps and social connection

Humans communicate with speech and the written word, but we communicate with maps, too. They're a basic—and profound—way to make sense of the world. More importantly, maps are a way we share our understanding of the world and are one of the key ways we establish and maintain professional and social ties. We tend not to think of them as social phenomena, but making and sharing maps is an act of communication—of connecting with people.

As we've learned, the hippocampus and entorhinal cortex are important to making sense of our location and movement through space and in sorting and filing away places and memories. But are the parts of the brain we use for spatial cognition and navigation engaged in the way we simply peruse a map? The research doesn't seem to have dealt with this question. But I can't help assuming that those grid cells and specialized arrays are actively firing as we gaze at a map and imagine—or remember—the landscapes they depict.

## Stories and the brain

Meanwhile, stories have their own impact on the human brain. In fact, listening to or reading character-driven stories literally changes our brain chemistry, according to Paul Zak, founding director of the Center for Neuroeconomics Studies at Claremont Graduate University. According to Zak, character-driven stories trigger the production of a neurochemical called oxytocin, putting us in a receptive, cooperative frame of mind. In a controlled study looking at charitable giving, Zak found that people experiencing the effects of oxytocin donated to 57 percent more of the causes, and gave 56 percent more money overall, than those not exposed to it.

As Zak indicated in the *Harvard Business Review* (2014), our bodies produce oxytocin when we feel trusted or kindly treated, and the chemical increases our empathy.

Do all stories produce floods of oxytocin? No. What elicits the responses Zak describes are stories that are character-driven and incorporate the sort of tension-and-resolution plot structure familiar to novelists.

The lesson is that stories in general, and online multimedia stories in particular, will likely be more effective if they feature a main character or hero. For example, the Amazon Conservation Team, a small nonprofit that works with Indigenous groups in the upper Amazon basin to help them protect their rain forest habitats, has featured individual tribal members in several of their multimedia stories. Similarly, scientists and staff in the field can be profiled as heroes as well. Scientists are trained to be dispassionate and often resist inserting themselves into narratives. But the customary objectivity found in academic journals is often self-defeating in a storytelling context, especially if stories are aimed at a general audience. People love to learn about people and are likely to empathize with a scientist—and better understand the relevance of their research—if their passion and motivation are incorporated into a narrative.

Sharon Kitchens, a member of the ArcGIS StoryMaps editorial team at Esri, worked with the Maine Coast Heritage Trust on a story called **Doing Conservation on the Ground** ⤤. She described how data gathered in the field is incorporated into the organization's GIS and used for planning and

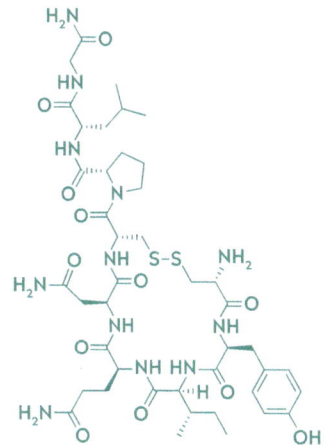

The oxytocin molecule.

*Maps and map creation are our bread and butter and a good two-way street between stewardship and land protection.*

Tatia Bauer, MCHT Regional Steward

As a regional steward Tatia works to ensure that a piece of land maintains a high ecological value while also serving as a resource to the local community, either as a place where one can go for walks or as a tract of scenic land that gives them a sense of place with their surroundings.

Tatia says geospatial technology is a part of everything she does. She recently submitted a grant application which required summarizing the various habitat types in the project area as well as rare and endangered species. Tatia used GIS to establish the project site boundaries and then compared them to existing databases and information she obtained from the State of Maine's Department of Inland Fisheries and Wildlife (through their Beginning with Habitat program).

"MCHT has been using ArcGIS software for so long now that it's difficult to imagine our work without it," shares MCHT Eastern Maine Land Protection Assistant Soren Denlinger. "ArcGIS allows us to access and display accurate, current data across the entire organization which can't be taken for granted."

Tatia Bauer standing in Forbes Pond Preserve, consists of 960 acres of forest and wetland just north of Prospect Harbor in the town of Gouldsboro.

When the organization is considering acquiring a property, Frank and Soren produce graphs and maps showing the ecological value of that parcel and its surroundings. That usually means including everything from endangered and threatened species to environmental issues such as future sea level rise. As different members of the organization meet to determine whether to move a project forward, Frank says GIS tools can make things really pop. "There is varied experience in regard to using GIS as an analytical tool, but using a map is not foreign to anybody," he explains. "They can look at a map and make some decisions based on the layers and spatial patterns depicted on that map."

Frank Cangelosi working in the field on Clark Island Preserve, St. George

A story about land conservation features photos of workers in the field.

management. But she also personalized the story with descriptions, quotes, and images of Trust staff working on site. The result is a much more engaging narrative that not only gives readers relevant facts but reveals the passion and commitment of conservation professionals to their work and to the natural world.

Heroes need not be human in all cases. Several conservation organizations have created multimedia stories featuring an individual migratory animal as a hero. In one collaboration, my team worked with the National Audubon Society to help raise awareness about migratory birds by creating the story **Aerial Odyssey: Bird Migration in the Americas** ↪, depicting a year in the life of Diego, a Swainson's hawk that was outfitted with a tracking device in Utah. We developed the story as a modest companion to a much more ambitious effort: Audubon's Bird Migration Explorer, a map-driven web resource that aggregates tracking data for hundreds of migratory bird species. In Diego's case, our hero flew north to Idaho for the summer before beginning his fall migration. The story followed him to his wintering grounds in the Pampas of Argentina, describing the hazards he faced along the way and profiling a Colombian biologist with a passion for raptors.

The development in recent years of ever-smaller tracking devices has brought to light the mind-boggling feats of many migratory songbirds. Only recently have we been able to follow the peregrinations of the blackpoll warbler, for instance. Many individuals breed in the Alaskan Arctic and migrate every fall to northern South America. Thousands of blackpolls pause in Nova Scotia on their southward trek and fatten up before taking off for the coast of South America, some flying nonstop for 72 hours. We all have at least a vague notion about the importance of protecting animals

Diego's return to Idaho marks the end of an odyssey that is both an act of individual endurance and representative of the annual pulsation of billions of birds, northward to breeding grounds and southward to non-breeding habitats. That these legions of migratory birds—weighing less than an ounce to just a few pounds—defy a gauntlet of natural and human-caused hazards to complete their seasonal movements is one of the great wonders of the natural world.

and their habitats. But telling stories about the dramatic life histories of migratory species, as revealed by remarkable technologies that include GPS and miniaturized electronic tracking devices, personalizes conservation and gives us a deeper appreciation of the feats of endurance that migrating animals accomplish.

## Left brain, right brain?

There's another thing to consider when contemplating maps, stories, and the brain: the notion of rational and quantitative versus intuitive and imaginative. It's an outgrowth of a theory, originated in the 1960s by Nobel Prize–winning psychobiologist Roger W. Sperry and since debunked. He posited that the lateral halves of our brains process information differently. The left tends to be logical, sequential, linear, and quantitative, while the right is more imaginative, holistic, intuitive, nonverbal, and creative. He argued that some of us are more left-brain dominant, while others are right-brain dominant.

   The left-right notion may not be valid, but the tension in our minds between rational and emotional is real. I like to think that maps tend to appeal to our more objective and rational traits, whereas the photographs and videos in multimedia stories stimulate the parts of our brains that react emotionally and aesthetically to stimuli (although a beautiful map likely engages both realms). At any rate, an advantage of multimedia storytelling is that it engages a richer array of mental processes than a map

Readers follow the movements of Diego, the avian hero of a story on bird migration.

In our story, **The Diversity of Life** ⊝→, a map appeals to our rational natures, and a photo of a frog stimulates our emotional, intuitive side.

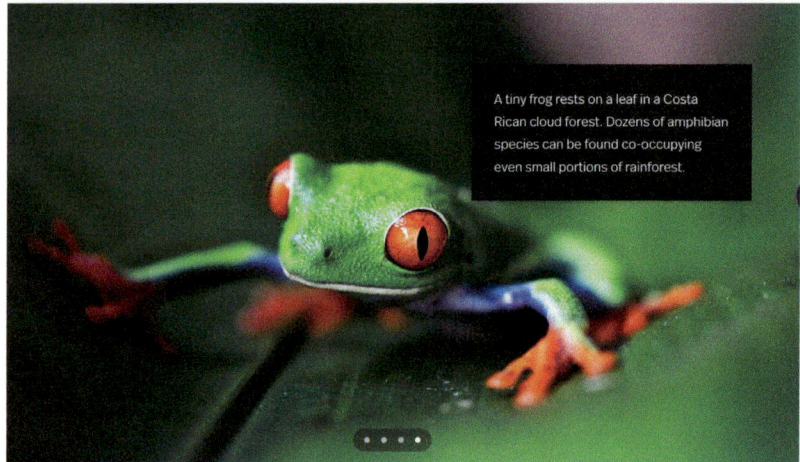

A tiny frog rests on a leaf in a Costa Rican cloud forest. Dozens of amphibian species can be found co-occupying even small portions of rainforest.

alone. I love maps but admit that they can be a bit distant and dispassionate. When did a map move you to tears? Images, however, can be deeply impactful. Pair maps with beautiful, evocative photos and videos, and you've got a powerful combination that is more likely to engage the whole brain.

Putting all these ideas together paints a picture of networks of neurons firing and neurochemicals activating across the length and breadth of our skull cavities as we engage with map-enriched stories:

- Maps stimulate rational thinking while also lighting up the more primitive hippocampus deep within our brain.
- Relatable characters and narrative arcs trigger neurochemical production, putting us in a cooperative state of mind.

- Visuals appeal to our intuition and aesthetic sense while eliciting deeply felt emotions.
- Grid cells, place cells, head direction cells, and border cells blink while we conjure up imagined images of landscapes or retrieve memories of our childhoods.

The scenario may be an exaggeration, but it might help explain my life-long interest in cartography and why I'm enthralled by maps.

A greater truth is that stories and maps tap into some of the basic things that make us human. As children we listen to stories at bedtime; as grand-parents we tell stories of bygone days. Maps? Like stories, maps help us understand. They help make sense of a dizzyingly complex reality. They bind together communities and cultures, and they bind us to our world.

# The Lines That Shape Our Cities

Connecting present-day environmental inequalities to redlining policies of the 1930s

**Authors:** Digital Scholarship Lab at the University of Richmond, the Science Museum of Virginia, and Esri

**Medium:** ArcGIS StoryMaps

**Story behind the story:** A New Deal–era agency called the Home Owners Loan Corporation (HOLC) compiled maps of dozens of US cities during the Great Depression. The maps used a rating system to categorize the desirability of neighborhoods; the judgments involved in drawing boundaries often reflected the racial and ethnic bias of the mapmakers—a phenomenon popularly called redlining.

**Why it's special:** *The Lines That Shape Our Cities* begins with a schematic representation that combines descriptive terms with color-coded areas (*right*). The story includes case studies from St. Louis, Missouri; Montgomery, Alabama; Fort Wayne, Indiana; and Oakland, California, illustrating how environmental problems and the impact of misguided redevelopment projects reflect the lingering legacy of redlining many decades after the maps were made.

The lines that shape our cities

Connecting present-day environmental inequalities to redlining policies of the 1930s

The story's title panel (*top*) and introductory map treatment (*above*) symbolize the classification of urban neighborhoods and their long-lasting impacts.

An Oakland neighborhood chosen for demolition as part of a redevelopment effort.

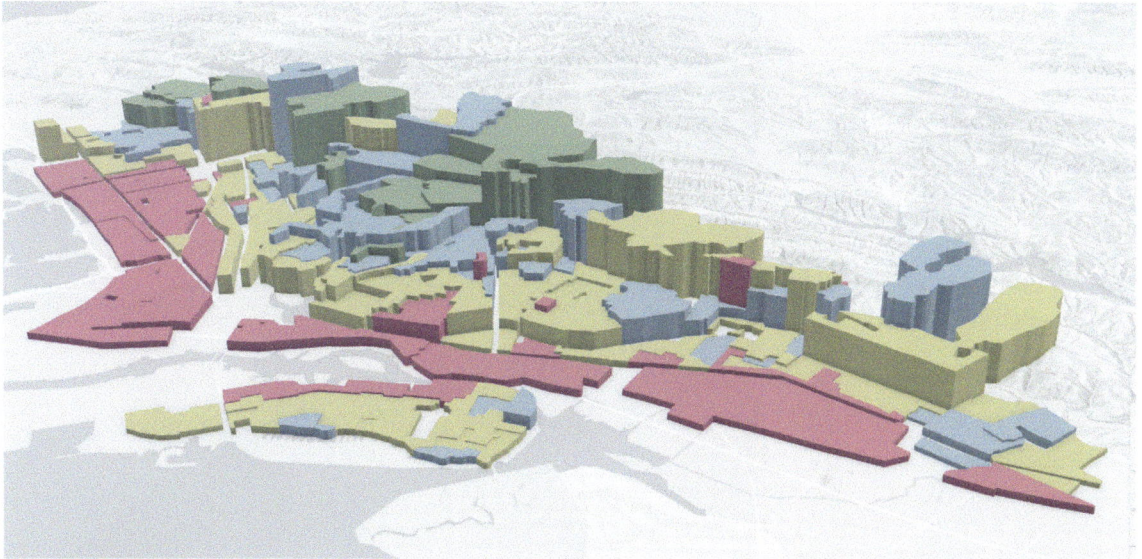

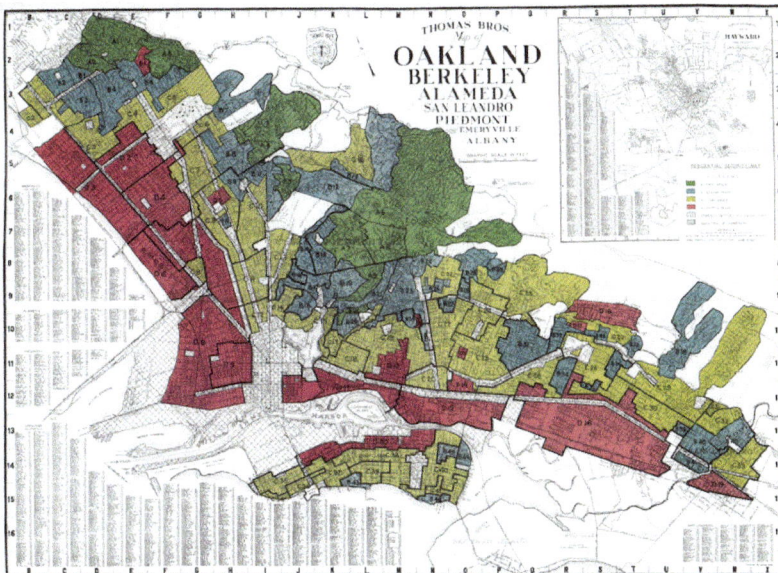

*Above*: A 3D representation showing the relationship of topography to HOLC categories. Neighborhoods at lower elevations received less favorable categorization.

*Left*: A HOLC map of Oakland and neighboring communities, graded from *A* (green, most desirable) to *D* (red, least desirable).

# The Old Man of the Mountain

In 3D

**Authors:** Matthew Maclay and team, Dartmouth College

**Medium:** ArcGIS StoryMaps

**Story behind the story:** On May 3, 2003, some 2,000 tons of granite tumbled from atop a cliff in Franconia Notch, New Hampshire. The incident might have created little notice but for the fact that it included a profile that, when viewed from a certain angle, looked like the head of an old man. The formation was iconic and was so well-known that it appeared on New Hampshire license plates.

**Why it's special:** Twenty years later, a team of researchers used archival photographs of the pre-collapse cliff to construct a 3D model of the formation, which they digitally reattached to the cliff face. The purpose of their study was to better understand bedrock weathering and rockfall in the Notch.

**Author:** Matthew's story uses 3D visualizations to zoom into the rockfall site and enable readers to peruse before-and-after images of the Old Man's demise.

*Top*: The digitally reconstructed rock formation. Shapes and textures were based on multiple historical photographs. *Bottom*: 3D model of the cliff face after the rockfall.

Photo: Workers installing turnbuckles in 1958. [1]

One of the archival photos on which the digital reconstruction was based.

The Old Man's famous profile before and after its disappearance.

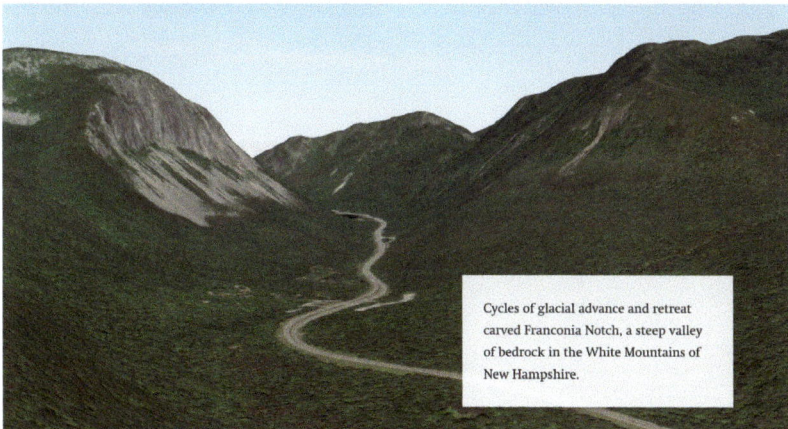

Cycles of glacial advance and retreat carved Franconia Notch, a steep valley of bedrock in the White Mountains of New Hampshire.

Web scene of Franconia Notch. The Old Man had graced the cliff face on the left.

NORTH KOREA

Pyongyang

Nampo

Hardened artillery sites

Pyongsan

Haeju

Ongjin

Goyang

Seoul

Incheon

Bucheon

Songnam

Suwon

Wo

# 4

## From Analog to Digital

Sokcho

60km artillery range

Gangneung

JTH KOREA

Population density (white and blue) mapped against hardened North Korean artillery sites and artillery range.

# The long road to digital at National Geographic

I began my career at the National Geographic Society in 1983 after working for several years as a freelance illustrator and designer. My initial position was assistant art director in the magazine's art department. Our primary mandate was to produce infographics and historical reconstructions for the magazine. I frequently rubbed elbows with photographers and picture editors, nearly all of whom worked with images captured on Kodachrome 64 color slide film. The magazine was entirely an analog enterprise. We created memos on electric typewriters and circulated them using metal carts wheeled down long corridors by clerical workers. The magazine maintained its publishing schedule in a ninth-floor control center, with article topics displayed on movable cards attached to large wall panels. Photographers spent months on assignment, with the aim of having perhaps a dozen of their images appear in the pages of the magazine.

As recently as the mid-1980s, visitors to the map division, housed on the third floor of the society's brand-new M Street building across a courtyard from the magazine's editorial offices in downtown Washington, DC, had plenty of cool stuff to see. Multiple flat file cabinets occupied much of the floor space. Among them large light tables glowed, over which skilled cartographers hovered. I summarized the analog map production process in the introduction to the *Atlas of the World*, seventh edition, published in 1999: "Cartographers carefully scribed roads, rivers, and other map information onto chemically treated plastic film, using a separate film sheet for each category—as many as 120 per map. Place-names were meticulously glued onto an additional layer. A photographic process combined these layers into sets of films. These were then delivered to the printer for the final preparation of the metal plates from which the *Atlas* was printed."

Computers had existed for many years, of

One of my early designs for *National Geographic Magazine*, for an article on Italy's "shadow economy."

course, but they were bulky, expensive, somewhat mysterious contraptions managed by specialists. I'm old enough, in other words, to have come of age in the latter years of the analog age and to have weathered the long, exciting, and sometimes painful transition from analog to digital.

By the mid-1980s, the society's transition to the digital age had begun. Its physical manifestation was a hulking machine that lurked behind glass panels in a refrigerated room in a corner of the map division. It was called the **Scitex Image Processing System**, manufactured by an Israeli company. Scitex equipment was originally developed to create and refine textile fabric designs. My friend and former colleague, Kevin Allen, joined the society in 1980 as a draftsman, doing traditional photomechanical work, but was trained early on in using the Scitex in concert with a film plotter and a scanner. As he described it, "You'd scan a map image; you'd use it in the background and hand-digitize elements, assigning layers to different categories of content (roads, borders, rivers, et cetera)." The society actually had two Scitex stations—the R-280 for digitizing and the R-300 for color rendering of artwork, primarily illustrations and shaded relief, to be combined digitally with linework and typography. The cartographic division employed four people as Scitex operators; they worked two shifts a day, with one person each on the 280 and the 300. Together, the workstations cost about a quarter-million dollars, and they were vastly slower and less powerful than the mobile devices we carry in our pockets and purses today.

During this early experimental period, the society also purchased equipment and software from Intergraph and from a little company called ESRI,

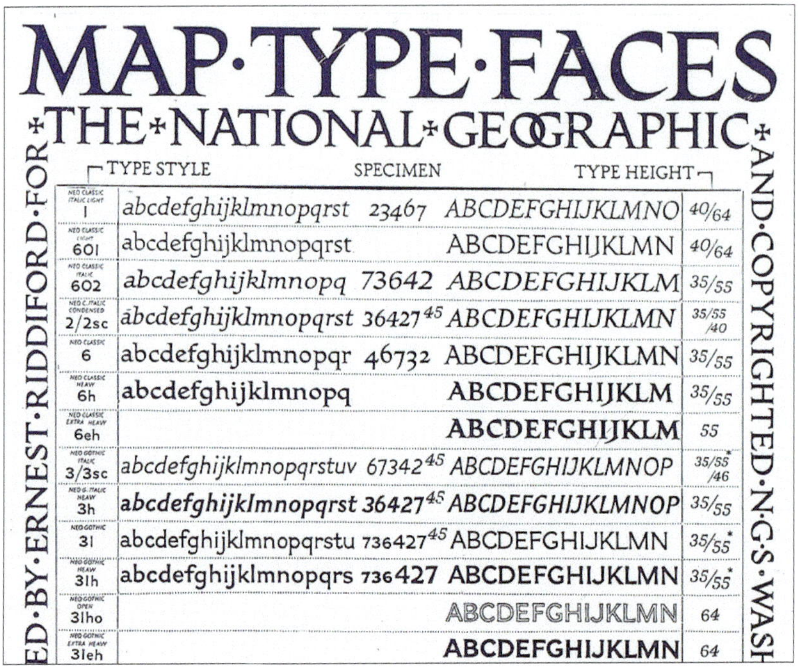

# MAP·TYPE·FACES
## ✠THE✠NATIONAL✠GEOGRAPHIC✠

Left margin (vertical): ED·BY·ERNEST·RIDDIFORD·FOR
Right margin (vertical): AND·COPYRIGHTED·N·G·S·WASH

| TYPE STYLE | SPECIMEN | TYPE HEIGHT |
|---|---|---|
| NEO CLASSIC ITALIC LIGHT 1 | abcdefghijklmnopqrst 23467    ABCDEFGHIJKLMNO | 40/64 |
| NEO CLASSIC LIGHT 601 | abcdefghijklmnopqrst    ABCDEFGHIJKLMN | 40/64 |
| NEO CLASSIC 602 | abcdefghijklmnopq 73642    ABCDEFGHIJKLM | 35/55 |
| NEO CLASSIC CONDENSED 2/2sc | abcdefghijklmnopqrst 36427$^{45}$    ABCDEFGHIJKLMN | 35/55 /40 |
| NEO CLASSIC 6 | abcdefghijklmnopqr 46732    ABCDEFGHIJKLMN | 35/55 |
| NEO CLASSIC HEAVY 6h | abcdefghijklmnopq    ABCDEFGHIJKLM | 35/55 |
| NEO CLASSIC EXTRA HEAVY 6eh |    ABCDEFGHIJKLM | 55 |
| NEO GOTHIC ITALIC 3/3sc | abcdefghijklmnopqrstuv 67342$^{45}$    ABCDEFGHIJKLMNOP | 35/55 /46 |
| NEO G·ITALIC HEAVY 3h | abcdefghijklmnopqrst 36427$^{45}$    ABCDEFGHIJKLMNOP | 35/55 |
| NEO GOTHIC 3I | abcdefghijklmnopqrstu 736427$^{45}$    ABCDEFGHIJKLMN | 35/55 |
| NEO GOTHIC HEAVY 3Ih | abcdefghijklmnopqrs 736427    ABCDEFGHIJKLMN | 35/55 |
| NEO GOTHIC OPEN 3Iho |    ABCDEFGHIJKLMN | 64 |
| NEO GOTHIC EXTRA HEAVY 3Ieh |    ABCDEFGHIJKLMN | 64 |

Charles Riddiford (*above*), and a sampling of the fonts he designed.

or Environmental Systems Research Institute. The Intergraph investment ultimately did not work for us, but the software, called ARC/INFO, was working its way into the cartographic division's production process and eventually demonstrated its utility many times over.

Another facet of the digital transformation was the digitization of National Geographic's cartographic type fonts. Back in the 1930s and 1940s, a skilled and stylish cartographer named Charles Riddiford designed a series of fonts tailored to specific roles in the society's maps, with different fonts for cities, towns, rivers, and other features. The fonts are proprietary and, along with the colorful "tint bands" that define international boundaries, are a primary reason that Nat Geo's maps are so instantly recognizable. The fonts were digitized during the 1980s and are still very much in use.

Placement of type on National Geographic maps was carefully considered. I remember stumbling upon a multipage document created by a reserved and meticulous cartographer named Tom Gray, a longtime member of the cartographic division staff who moonlighted as a member of a locally renowned Bluegrass band called the Seldom Seen. His document delineated the multiple ways type should be placed relative to a town spot or other map feature. The document listed and described options by order of preference. Labels had to be a certain distance from the features they named; placement depended in large part on how dense the feature's

TOWNS

*Tom Gray*

The distance from a townspot to the name should be between ½ and 1 width of a number one townspot.

Following is a list of 40 possible ways to place a name to a townspot in order of preference. These are not the only positions at all; names can and should be placed in positions between those shown here. This list just illustrates the priorities to be kept in mind when naming a townspot or other small map symbol.

1. bottom of spot tangent to bottom of name on right
2. top of spot tangent to top of name on right
3. bottom of spot tangent to bottom of name on left
4. top of spot tangent to top of name on left
5. top of spot tangent to top of single ascender on left
6. bottom of spot tangent to bottom of single descender on left
7. name centered right of spot
8. name centered left of spot
9. above and to right (45 degrees is best angle from spot)
10. below and to right ( . "    "    "    "    "    "    " )
11. above and to left ( "    "    "    "    "    "    " )

*excellent* / *very good*

A fragment of Tom Gray's map label placement guide.

*Below*: This detail from the society's *World Atlas*, seventh edition, shows how place-names were meticulously adjusted and curved to enable maximum detail in densely populated areas.

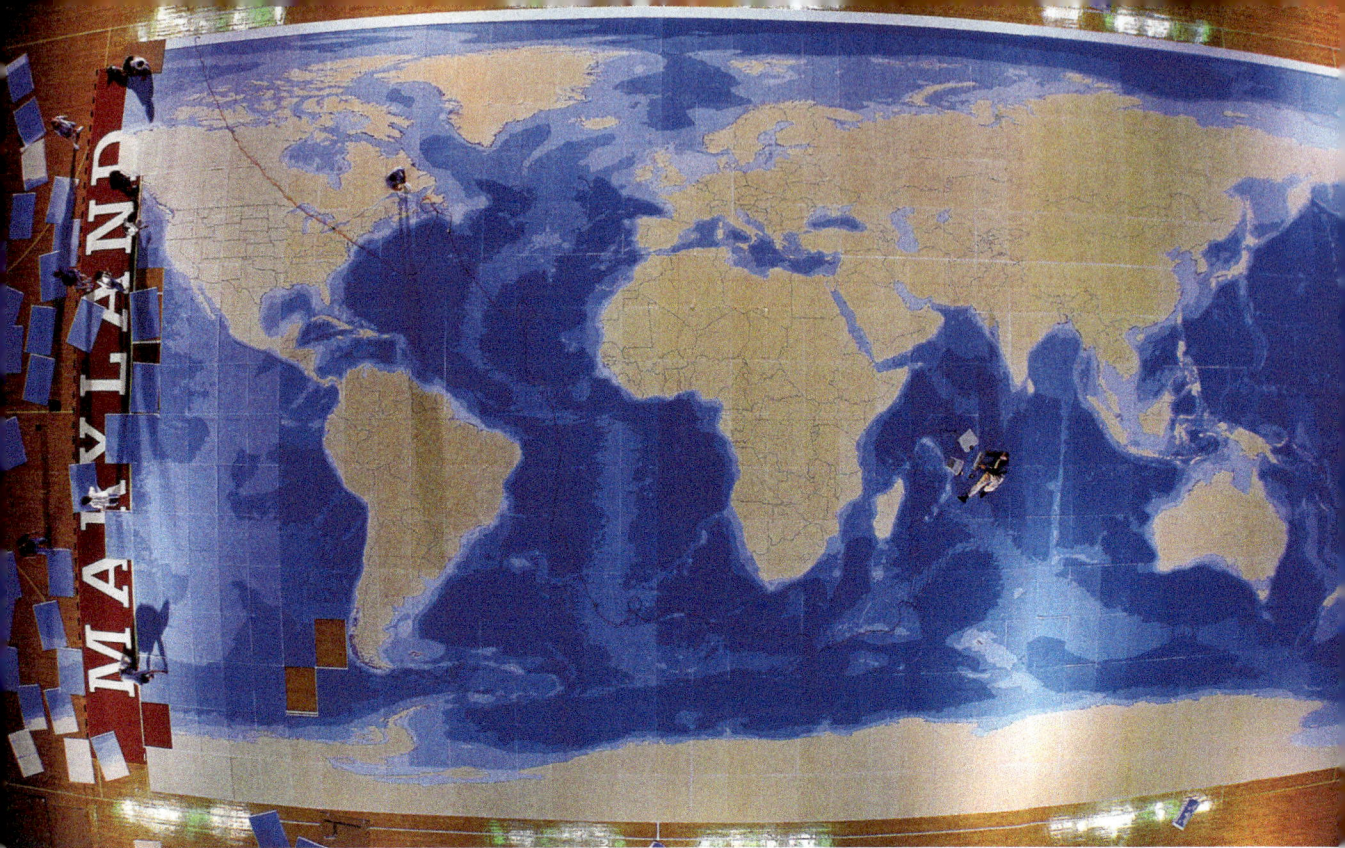

Printouts of the 1:1 million scale Digital Chart of the World were assembled as a visual stunt using the University of Maryland's basketball court.

neighborhood was with other items. Type might be curved but had to be curved in certain ways.

The long, painstaking process of transforming National Geographic's maps into the digital realm was partially facilitated by the use of a seamless map of the world compiled by the Defense Mapping Agency (now known as the National Geospatial-Intelligence Agency). With the help of outside contractors, high-resolution scans were made of the maps—or plates—within the society's *World Atlas*, which represented world regions at a variety of scales and projections. Using ARC/INFO, portions of the Defense Mapping Agency's **Digital Chart of the World (DCW)** were cut out and reprojected to match the scale and alignment of the *World Atlas* plates. This second scanned layer was used as a reference in combination with the *World Atlas* scans to maximize the spatial accuracy of the new digital data.

As a demonstration of the scale of the 1:1 million map, the society assembled an array of large-format printouts of the DCW on the floor of the University of Maryland's field house and photographed them for the *World Atlas*. All of this occurred long before anyone imagined Google and Apple Maps.

As computer technology rapidly evolved, the division transitioned to a two-step publishing process. In the first step, digital GIS databases were created using ARC/INFO technology. A key advantage of the use of GIS was that all 140,000 place-names in the *World Atlas* resided in a single table. If a place-name were to change, all of that place's appearances within the *World Atlas* plates could be identified and quickly updated, saving countless hours of manual effort and simplifying the creation of the atlas index when going to press with updated versions.. In the second step, those files would be exported to Macintosh-based desktop graphics software (primarily Adobe Illustrator and Photoshop), where cartography would be refined and combined with artwork, text, and other elements to create press-ready files.

The first generation of National Geographic's digital cartography existed in the form of separate databases, one for each of its *World Atlas* plates. A later step was to combine those digital plates into a nested set of seamless cartographic databases of the entire world at a variety of scales. Those databases could then serve as the basis for many maps—primarily the large-format supplement maps folded into the magazine, as well as page maps that appeared within magazine articles.

This brief description belies the massive scope of the digitization process, requiring countless hours of painstaking manual work. It was a huge investment, but the payoff in increased productivity and flexibility was massive. When I first worked in the cartographic division, leading a small group of designers, its numbers exceeded 100 people. When I left National Geographic to join Esri in late 2010, some 27 people were producing a considerably larger volume of work and were able to adjust to a changing world with mouse clicks and keystrokes applied to databases rather than hand-drawing corrections to dozens of pieces of film.

I've described the predigital cartographic division's drafting tables, flat files, and specialized equipment that gave the division's offices such a distinctive look. Visit a cartographic shop today, and what you see will look much like any corporate workspace, with the exception of a large-format plotter or two and the visuals—maps instead of spreadsheets—that appear on computer screens. But the work that goes on in these shops is more exciting than ever. It was a huge honor for me to have helped maintain and, in modest ways, enhance National Geographic's tradition of superb cartography. I confess to feeling somewhat undeserving of this privilege: What little cartographic expertise I had was learned on the job. I have had no formal training in cartography, geography, or GIS. I approach maps from a design and editorial viewpoint and with the same visceral love of their beauty—and sense of wonder at the world they represent—that I felt as a child.

A schematic illustrates the concept of multiscale seamless map databases.

# Storytelling with maps on paper and screen

I've described how digital technology revolutionized the process of preparing maps for printing on paper—digital production of analog maps, in other words. A more profound impact of the digital revolution was the explosion of digital media, involving the creation and publication of maps for consumption not on paper but on the displays of computers, tablets, and mobile devices. This aspect of the revolution thrust the ancient art of cartography into dramatically different realms. Some of the time-honored conventions of mapmaking survived this revolution intact, or nearly so. But the fundamentally different experience of consuming maps on digital devices mandated an array of updates to cartographic conventions. It's an interesting mental exercise to catalog the differences between the analog and digital universes.

Maps in both media can be highly effective storytelling devices, and they inevitably bear similarities. In fact, some time-honored cartographic principles having to do with scale, projection, typography, and symbolization apply equally in analog and digital realms. But print and web are dramatically different media, and the techniques of mapmaking—and storytelling—can be strikingly different because of the constraints and advantages offered by each.

I've been fortunate to learn a bit about these principles and practices during my time as National Geographic's chief cartographer and in my more recent incarnation as the founder of the Esri ArcGIS StoryMaps team. My career at Nat Geo began when the internet was barely more than a theoretical concept, so I've lived through the long and sometimes painful transition to the digital age. It's been exciting to witness this essential shift from within the worlds of cartography and storytelling and to learn and adapt as the transition occurred.

The following section describes a few of the key differences between analog media that have all sorts of implications for mapmaking and storytelling.

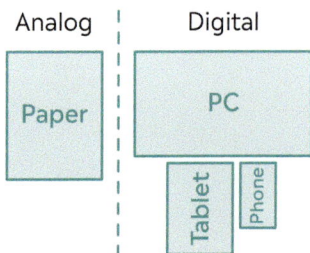

## Size

The one constant that designers took for granted in the analog age was that paper size, once determined, would remain fixed. Considerable size and proportion differences among digital devices mean that stories in digital media should be responsive: They need to adjust to accommodate screens and aspect ratios. The ArcGIS StoryMaps team strives to make the stories created by its users fully and pleasantly readable across computers, tablets, and phones. That means that sometimes a story's elements are rearranged, or images cropped, based on dictates of screen size—especially for mobile devices—where screens are small, and readers might be viewing in both portrait and landscape modes.

## Motion

Digital media can be, and often is, dynamic. It can move, update, and respond to a viewer's actions. Analog media is static. It just sits there. You can carry the book or magazine around, but the stuff inside it will stay the same.

Motion is key to digital-age storytelling. Readers navigate stories by scrolling, clicking, tapping, swiping, and pressing arrow keys. Maps pan and zoom; maps in series reveal change over time; map animations bob and weave. The availability of motion has created vast new opportunities for cartography.

A word of caution, though—there is such a thing as gratuitous motion. Things that arbitrarily bounce, spin, or jiggle can get on a reader's nerves. In the 3D realm, excessive flying transitions from one location to the next, especially if altitude and orientation change midflight, can be nauseating. Motion should thus be employed tactically.

Examples of excessive motion reside within many stories that use 3D visualizations. Google Earth used to feature hundreds of virtual tours that "fly" readers from place to place across a 3D globe. The incessant whooshing from place to place quickly becomes irritating. I indulged in 3D overkill for a story I created using ArcGIS StoryMaps called **Peaks and Valleys** ⟲, wherein I took people on a continent-by-continent tour of our planet's highest, lowest, and most vertiginous locales. It's fun to spin from

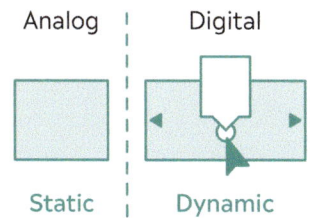

| Analog | | Digital |
|:---:|:---:|:---:|
| Static | | Dynamic |

*Peaks and Valleys* story.

Everest to K2 to Kangchenjunga and beyond—until your stomach starts to churn.

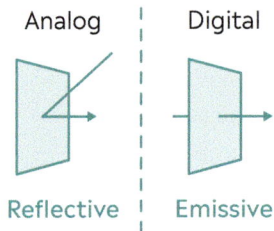

## Light

Analog | Digital

Reflective | Emissive

Traditional media uses reflected light; ambient light bounces off the paper into the reader's eye. New media uses screens that *emit* light, which has subtle but significant impacts on the viewing experience.

Light, or the difference between reflective and emissive, may be one of the reasons that attention spans on digital media tend to be shorter than on analog. We seem to read for shorter periods and less attentively on screens than on paper. Is it a generational thing? Younger "digital natives" are more comfortable reading on screens than older "digital immigrants." But Ferris Jabr argues in a 2013 *Scientific American* essay that eyestrain may result from prolonged reading on some screens or monitors. That's one of several reasons the ArcGIS StoryMaps team strives for brevity in our storytelling.

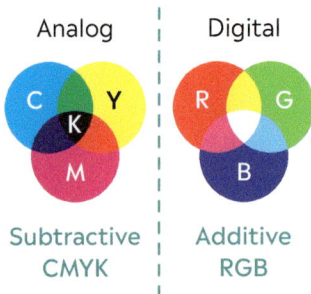

## Color

Analog | Digital

C Y
K
M

R G
B

Subtractive CMYK | Additive RGB

Print media typically uses four colors of ink: cyan, magenta, yellow, and black (CMYK). They're *subtractive*, meaning that inks absorb parts of the spectrum to produce color. Electronic screens are *additive*, using light instead of pigment. Red, green, and blue (RGB) light combined make white.

The range of hues available to us on screens is greater than on paper. In other words, inks have a more limited gamut of colors than light-emitting screens. Deep royal blues and brilliant yellow-greens, for instance, can only be approximated with standard printer's inks. That larger digital gamut is good news for web designers. But more vibrance isn't always better. Subtler hues are easier on the eye and mind and tend not to distract from the message you're conveying.

Hue

Value

Saturation

Whole books could be written about maps and color—and they have been. Cynthia Brewer's "color brewer" publications (including *Designing Better Maps: A Guide for GIS Users* [2024]) provide inspiration and practical advice for cartographers. I'll limit my pontification here to a couple of items. One has to do with the trio of characteristics that comprise color: **hue** (where a color sits on the rainbow spectrum), **value** (how pale or dark a color is), and **saturation** (which might be described as how clean or dirty a color is). I've called saturation the cartographer's secret because it can be so helpful in making some map elements recede into the background and others stand out.

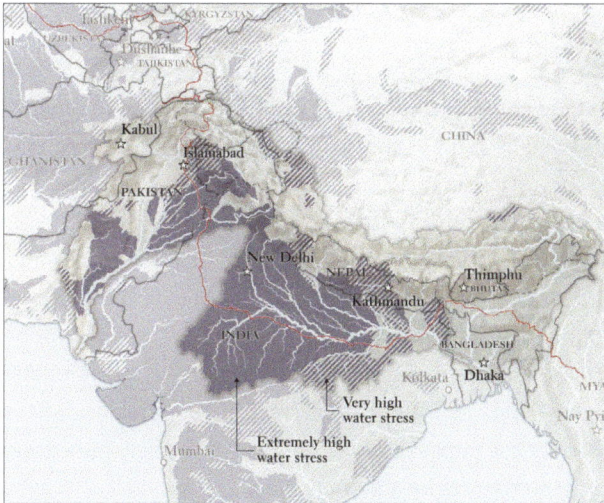

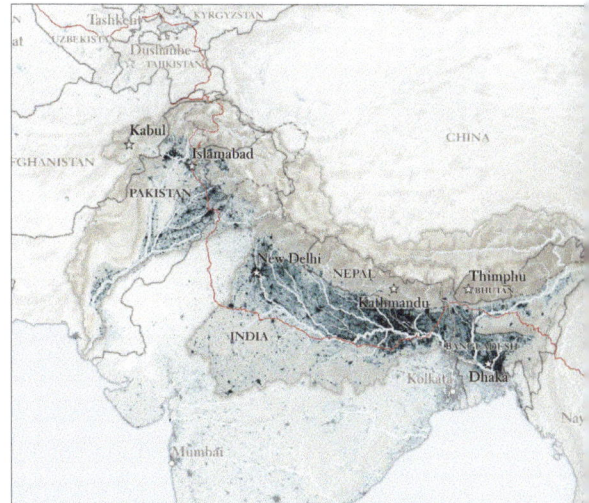

A fine example can be found in the maps that cartographer Cooper Thomas produced for **The River Roads of India** ⊖⊃, a story published by my team in collaboration with journalist Paul Salopek, who is on a years-long trek tracing the human diaspora from Africa across the world. Cooper chose a basemap of subtle grays and earth tones, and then positioned thematic information in somewhat brighter colors on top. But the colors he chose aren't fully saturated; they're toned back to help create a harmonious whole. Paul's route, however, is shown as a thin line in bright red. Within the story the two maps are neighbors in the narrative; as readers scroll, the first fades seamlessly into the second, allowing for easy comparison.

Continuing our paper versus screen, analog versus digital comparison...

Adroit color treatments utilizing hue, value, and saturation.

## Publishing

In print, publishing happens once and is essentially irrevocable. The printed item is more or less a permanent artifact. Web-based digital media, on the other hand, can be republished again and again. Edit your work, refresh the screen, and the story has new life. Publishing, formerly a painstaking, protracted process, is largely instantaneous on the web.

Gone are the days when I would put the finishing touches on a map, wait for weeks (even months!) to finally see it in print, and then immediately spot a design or editorial element that I should have approached differently. The same regrets occur for web-based efforts, but a few minutes of work, and a click of the Publish button, will usually resolve the issue and soothe the heart. For those of us who are veterans of the analog age, this ability to instantly and repeatedly update is a continuing miracle.

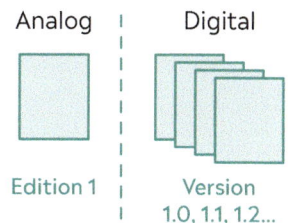

Analog | Digital

Edition 1 | Version 1.0, 1.1, 1.2...

Analog | Digital

## Permanence

Finally, permanence is an issue that we all should consider. Will my future great-granddaughter be able to go to a library and find the printed maps I designed for National Geographic? Probably. Will she be able to dig up my ArcGIS StoryMaps stories? I doubt it.

Many of us tend to assume, consciously or otherwise, that print on paper is ephemeral and that the ones and zeroes of digital files are permanent. Many of us, myself included, invest time in scanning our older family snap-shots to preserve or archive them—and to jettison shelf-consuming shoe-boxes full of old prints. I have three separate copies of my family archives, which makes me reasonably confident that my digital files are permanent. But in the long term, it's the paper that's more likely to survive. A century or two from now, will our devices be able to parse JPEGs and PNGs? Maybe. Will my paper prints persist? Barring flood, fire, or overzealous houseclean-ing, yes.

## Two Koreas, two mediums

To more clearly understand the similarities and differences between print and digital storytelling, it might be useful to study an actual example of print versus digital approaches to the same topic. I had the pleasure of working on two map-based accounts of the Korean conflict, one in print for National Geographic in 2003 and a second in 2017 after I joined Esri. Let's look at the two projects, both called **The Two Koreas** ⊂⊃.

For the 2003 effort, I collaborated with cartographers and writers at National Geographic to produce a large-format supplement to *National Geographic Magazine* on the Korean peninsula. One side was a detailed map of today's North and South Korea; the other featured a brief history of the Korean conflict, which occurred a half century prior to the map's publication.

Telling the complex, tragic story of a three-year-long conflict on a single sheet of paper posed some significant challenges. We developed a graphic style to make the large amount of information attractive and digestible. We chose a limited color palette, which we employed consistently across the series of maps. We incorporated a distinctive type treatment, added a torn-paper effect to archival photographs, and used a neutral background tint to help tie the many elements into a single composition.

There are dozens of items on the sheet, including photos, maps, text blocks, and titles, all visible at once and all vying for the reader's attention.

*The Two Koreas*: map supplement to *National Geographic Magazine*, July 2003.

Cover panel of the folded supplement map.

We used large numbers to guide readers through the sequence of episodes showing the shifting tide of the conflict. We featured one map—representing the maximum advance of the United Nations counterattack—as the centerpiece. Without a central focus, the design would have looked less like a poster and more like a grab bag of elements competing for center stage.

Another challenge posed by supplement maps was that they were folded for insertion into the magazine, requiring us to design a title panel that also aligns well with, and complements, the composition of the whole sheet.

Fast-forward to 2017, a time when tensions between the United States and North Korea were on the rise. My new team at Esri decided to produce a multimedia story about Korea's history, economy, and, for the North, nuclear ambitions. In much the same way that the Nat Geo team developed a visual vocabulary, we chose a red and blue-gray treatment that we used throughout the story. Not surprisingly, red represents North Korea; blue-gray, South Korea.

We used this split-color treatment for infographics, on photos, and for the maps themselves. And we opted for a black background, again as a unifying element and to add a bit of drama to a somewhat difficult topic.

No need, this time around, for a numbered sequence. Readers simply scroll through the narrative; although the story header includes bookmarks, a mouse-click can jump to various parts of the story.

The most striking difference between the old-media and new-media treatments is the map presentation. As a reader scrolls through the story, a historical map sequence reveals the control of territory seesawing back and forth between Communist and Allied forces.

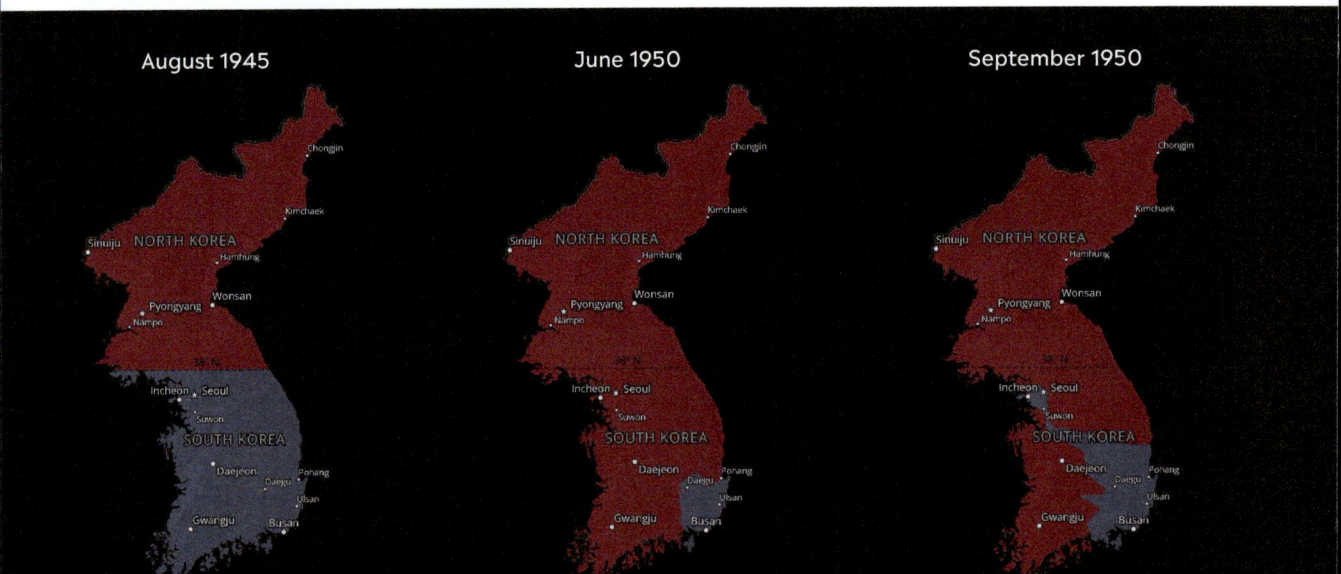

*The Two Koreas* multimedia story: cover and color palette.

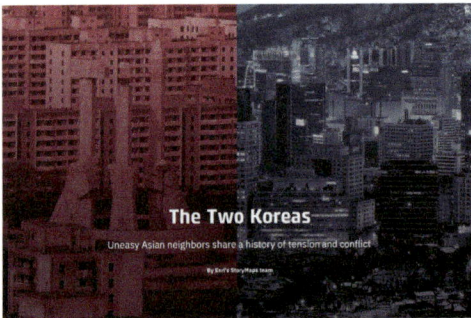

| NORTH KOREA | | | SOUTH KOREA |
|---|---|---|---|
| | ff4c4c | b6c9d6 | |
| | ff0000 | 96b2c5 | |
| | da0000 | 7a9db5 | |
| | bb0000 | 5a85a1 | |
| | 9e0000 | 47697f | |
| | 830000 | 375163 | |
| | 5e0000 | 283b47 | |

August 1945        June 1950        September 1950

This ability to see a map (or series of maps) change within the same space is arguably more effective than glancing back and forth among maps positioned across a sheet of paper, as shown below. It provides a visceral sense of movement and an ease of comparison that spatially separate maps just can't match. I can't show you, in the pages of this book, how this scroll-driven sequence works. And that's exactly the point: We're taking advantage of the digital medium to create effects that suit its context and capabilities, which can't be achieved on paper.

Having maps change as readers scroll is what we call **map choreography**. We frequently employ this technique and urge others to use it. In the case of ArcGIS StoryMaps, it's relatively easy to implement within its side-car and slideshow immersives. These "carto-dances" can include either

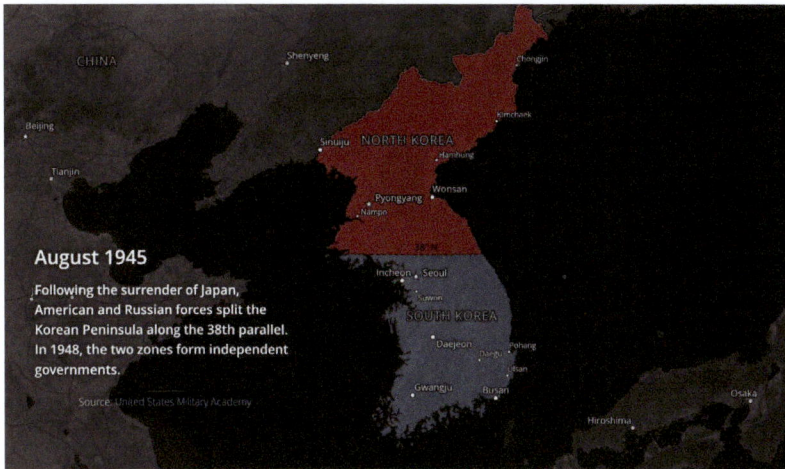

**August 1945**

Following the surrender of Japan, American and Russian forces split the Korean Peninsula along the 38th parallel. In 1948, the two zones form independent governments.

Source: United States Military Academy

An immersive section of *The Two Koreas* multimedia narrative (*left*) features a series of maps (*below*) charting control of territory over time.

October 1950

December 1950

1951–1953

sequentially changing the theme of the map as the scale and map extent remain constant or changing the zoom level and extent of the map as readers scroll, or both.

You'll notice that these maps aren't truly interactive. As much as I might wax ecstatic about panning, zooming, clicking, and swiping, my team and I have actually employed fewer fully interactive maps in our more recent stories.

We can use static maps to present only what the story needs, without the distraction of map interactions. Static maps load in a browser more quickly than dynamic maps (they're just an image, after all). And we can use the standard scrolling action as a kind of simulated interactivity. Readers can move up and down repeatedly to study a series of maps—meaning that they're interacting with the map series—in the same way that they're interacting with the whole narrative. They're not distracted by panning, zooming, or clicking. There's no interruption of the flow, no new functionality to learn.

We can also opt for a hybrid approach, presenting a dynamic map with pan and zoom controls while choreographing it to change as the reader scrolls. Most readers will ignore the map controls, but the minority who might want to explore the map can interrupt the narrative to pan and zoom. We'll further explore map choreography in chapter 8.

Both of the *Two Koreas* projects were successful, but the digital version gets my vote as the more effective of the two. Its mix of infographics and maps, tied together by a consistent and visually striking design style, taps the power of digital media to good effect. But it's a tough call: Each medium has its own strengths and weaknesses.

The printed map's "Forgotten War" narrative was similar in approach not only to the Korea supplement but to many of National Geographic's thematic supplement maps from that era. Looking back, Nat Geo's use of maps, text, artwork, graphs, and sometimes photographs seems like a precursor to the web-based multimedia storytelling techniques we use in creating ArcGIS StoryMaps narratives.

## Print and digital: Luxuries and limits

Despite the advent of the digital age, large-format printed maps still linger on the walls of countless classrooms, offices, and kids' bedrooms. These maps present an enormous amount of detail, sometimes including several thousand place-names, simultaneously. And in many cases, they do so with aesthetic panache.

These instantaneous, sweeping geographic panoramas are impossible on even the largest computer screens and retinal displays. I occasionally pine for the unique and exciting challenges of designing large-format wall maps. I was fortunate to have the opportunity to design many supplement maps for National Geographic back when budgets allowed for four or more inserted maps every year. We also produced—and National Geographic continues to create—beautiful wall maps of the world, continents, regions, and countries in a variety of styles and themes.

The trick with large-format printed maps is to guide readers through

Detail of a political map of the Korean peninsula, showing the cartographer's subtle storytelling techniques.

Maps unfold in layers of meaning.

**Larry Orman**
Founder and former executive director of Green Info Network

a subtle narrative that's the result of careful cartographic design choices. On a single sheet, there's no way to invoke timing or sequence as you can with web pages or mobile apps, where maps can be presented in succession. Even conventional political maps are designed to present information that "unfolds in layers of meaning," as my friend and longtime mapmaker Larry Orman likes to say. Typographic treatments are hierarchical, with continents and countries more prominent than cities, and large cities more visible than small towns. International boundaries are represented with tint bands, and terrain—less important on a political map—is shown in subtle shades of gray.

The detail on the previous page from the National Geographic Korea supplement illustrates these and other layers-of-meaning techniques. I've provided this magnified view to force a closer look at the cartographers' subtle storytelling strategies. Color is used to differentiate categories of information: blue for water features and red for transportation. Physical features such as mountain ranges are represented with curved and letter-spaced italic type.

Digital maps suffer the confines of small screens. Smartphones are puny in comparison with wall maps, and even the largest monitors fall short in size and resolution compared with the fine detail of printed wall maps.

The compensating factor for web and mobile is the ability to access multiscale maps that increase in detail as users zoom in to the map. As we all know, it's possible on Esri basemaps, Google Maps, and other web mapping apps to zoom almost instantly from a view of the entire globe to a map of a street corner at a scale where one inch equals 20 feet. A printed world map at that scale would be more than 100 miles across!

These online "slippy" maps, that invite users to pan instantly across landscapes and zoom effortlessly from neighborhood to planetary scales, offer power and insight that were unimaginable only a few decades ago. A relatively unsung impact of these multiscale, interactive maps—especially those that incorporate satellite imagery—is that they have subtly changed the way nearly all of us perceive the world. The link from driveway to whole Earth is ubiquitous and undeniable. The Apollo program's famous *Earthrise* photograph was a revelation; interactive maps abetted that planetary vision with a constant reminder that our locales, no matter how provincial or isolated, are part of a global system, a shared, ever-changing, and finite existence.

Is one medium superior to the other? Are we better off returning to the analog days or, at any rate, returning our focus to the art of print

The Apollo 8 famous *Earthrise* photograph.

cartography? The obvious answer is that both mediums are powerful and effective—they're just effective in different ways. Print is far from obsolete, of course; we can revel that we have two cartographic mediums in which to tell our stories.

If I were to distill into one idea the difference between map-based storytelling in print and digital, it would be that print provides the luxury of *space*—of presenting lots of fine detail across the expanse of a sheet of paper—while digital gives us the advantage of *time*, enabling us to unfold our narratives along the razor-thin membrane between past and future.

# Battles of the Civil War

## A timeline of tragedy and emancipation, 1861–1865

**Author:** American Battlefield Trust and Esri

**Medium:** ArcGIS StoryMaps and a custom multimedia story

**Story behind the story:** The ArcGIS StoryMaps team collaborated with the American Battlefield Trust to create two resources featuring chronologies of the battles of the Civil War. An ArcGIS StoryMaps story spotlights the war's biggest battles; a custom timeline app provides a more comprehensive listing.

**Why it's special:** Readers can click map symbols to access descriptions; panning and zooming on the map works as a filter, featuring battles within the current map extent.

**The author:** The American Battlefield Trust works to preserve America's hallowed battlegrounds and to teach the public about what happened there and why it matters. Battle descriptions and images were provided by the organization. The timeline perennially ranks near the top of the list of most-visited team-produced stories.

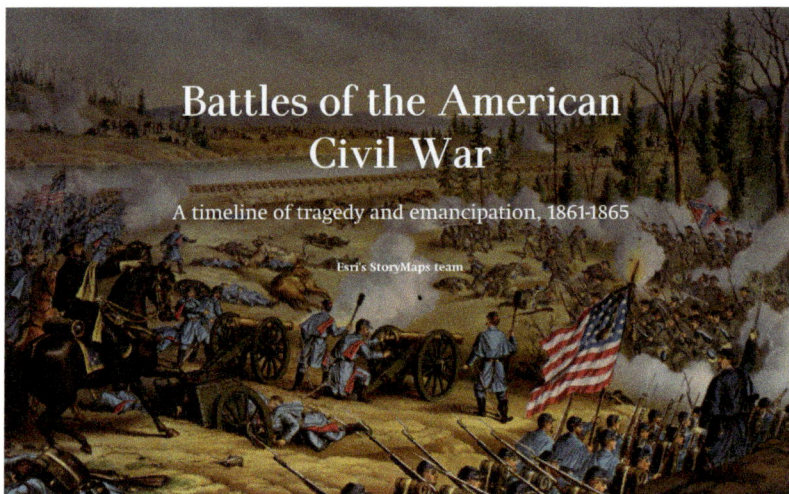

*Top:* Cover art and title. *Above:* Aftermath of Battle of Antietam, the single bloodiest day of the Civil War.

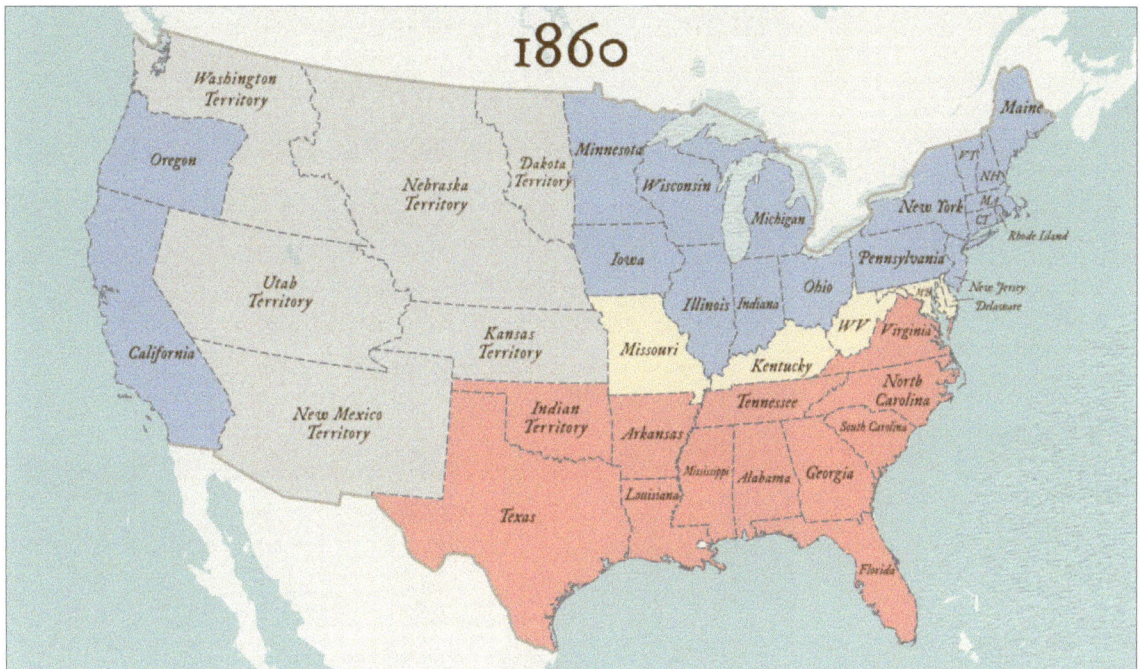

*Top*: An animated timeline shows cumulative battles through the Battle of Gettysburg, July 4, 1863. Red diamonds are Confederate victories, blue diamonds are Union victories, and yellow diamonds are inconclusive.

*Above*: Frame from an animated map showing Union (blue), Confederate (red), and neutral (yellow) states and territories.

# Travels with Godzilla

## A western birding adventure

**Author:** Allen Carroll

**Medium:** StoryMaps.com

**Story behind the story:** In the summer of 1971, I traveled more than 14,000 miles around the western US and Canada with Sam Test, a high school friend and fellow bird enthusiast.

A biology teacher had unexpectedly ignited a passion for birds in both of us; we indulged that passion with a summer of driving, camping, and spotting birds. We named Sam's underpowered Ford Maverick "Godzilla" and returned to our hometown of Indianapolis in late August with dozens of new species and a host of memories.

**Why it's special:** A half century later, I dusted off the journal I had kept and published it as a collection of stories.

Travels with Godzilla: A western birding adventure

The summer explorers with Godzilla (*top*) and camping gear (*bottom*).

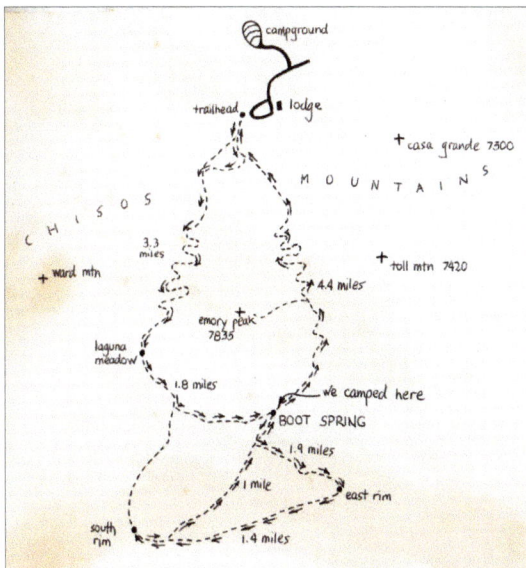

*Top*: Map of a driving and birding marathon. *Left*: Hand-drawn map of an overnight backpacking trip in Big Bend National Park, Texas. *Above*: Fire lookout in Umpqua National Forest, Oregon, summer home of a couple of college friends.

# 5

## The Journey
## to Storytelling

# Prelude: National Geographic

I was fortunate to work for 27 years at an organization that was all about storytelling, and for which maps were entangled within its DNA. As I mentioned earlier, I began my career at the National Geographic Society in 1983 as assistant art director on the magazine staff. I had freelanced for several years before that as an illustrator and designer, doing work for Johns Hopkins University, *The New Republic, The Washington Post,* and other Washington-area organizations. I loved my work, and the variety and freedom that self-employment offered, and I could imagine only one job that I might find preferable—and that was working for National Geographic. I had made a couple of unsuccessful attempts to get my foot in the door by showing my portfolio to an assistant in the magazine's art department. Finally, my work for the *Post* caught the eye of art director Howard Paine, who was to become my mentor. He offered me my dream job.

Cover and spreads from the *Historical Atlas of the United States,* published in 1988 and now out of print.

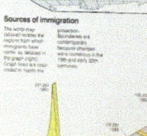
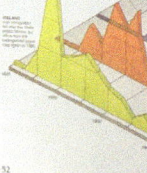

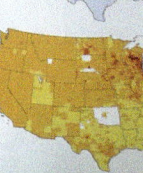

Although my home base was in the art department, primarily working on infographics and historical reconstructions, in the succeeding years I did two separate stints with the society's cartographic division before finally becoming chief cartographer during my last decade or so at National Geographic. First was a two-year effort to design and art direct a *Historical Atlas of the United States*, published in the society's centennial year of 1988. I was embedded amid mapmakers, and the posting served as intensive cartographic basic training. The atlas was an elaborate production, with scores of newly produced maps placed amid a panoply of images, artwork, timelines, and historical maps. It was old-media map-based storytelling of the most lavish sort. Sadly, the giant-format atlas is long since out of print, although it remains one of my proudest achievements. (There is a more recent National Geographic book with the same title, but it's an entirely different, and much less ambitious, effort.)

I returned to the art department at about the time the magazine's circulation peaked at nearly 12 million and as it neared its centennial in 1988. It was near the end of an era: None of us had any real notion that enormous changes in the media marketplace, brought on largely by the digital revolution and the advent of the internet, would foment a long, painful, inevitable decline in the magazine's reach.

Stint number two was to lead the cartographic division's small team of designers. We worked on several large-format map supplements to the magazine and designed countless "page maps" that accompanied magazine articles.

When Howard Paine retired, I again returned to the art department as its director and as a senior assistant editor for *National Geographic Magazine*, a fringe benefit of which was membership on its planning council. It was a thrill to play a small role in setting the magazine's direction. At this point, I figured I had achieved professional nirvana—helping foster one of the world's most revered brands, working with some of the most talented illustrators alive, rubbing elbows with eminent scientists, photographers, and explorers.

And yet, putting out one magazine issue after another started to feel a little repetitive, especially when some of the same article topics started to come around for a second or third time.

Maps had gotten under my skin. Yes, cartography was being held in less regard within the society's hallways. Yes, maps are a specialized and, in some ways, a limiting medium. But maps were indeed part of the society's DNA. The map division offered an opportunity to inherit a rich tradition of producing world-renowned maps and atlases. And it offered a degree of autonomy to pursue new opportunities—to *invent* new opportunities—as the organization struggled with declining circulation and a changing media environment. So when a position became available to lead the cartographic division, I jumped at the chance.

I returned to the division as it gained a new name, National Geographic Maps, and became part of National Geographic Ventures, a newly formed for-profit subsidiary of the nonprofit society. The launch of Ventures was the first of several rounds of wrenching changes as the society's leadership struggled to adapt to seismic shifts in the media marketplace and the continued decline of the magazine's readership. A cadre of cartographers continued to produce maps for the magazine while also finalizing the first acquisition in the society's history—a small, Colorado-based recreational mapping company called Trails Illustrated. We also formed an alliance with a Pennsylvania group called GeoSystems Global Corporation, later to become MapQuest. With GeoSystems, we published a Nat Geo–branded road atlas and a series of travel maps.

You can still buy the road atlas, by the way; it continues to be updated periodically (even though the cover hasn't changed since I designed it some 15 years ago) and serves a shrinking audience that longs for maps more tangible than the ones Google and Apple put on our smartphones. Its beautiful maps were designed by British expat Andrew Skinner, who left GeoSystems and joined Esri to create gorgeous multiscale digital basemaps for its online content service.

Meanwhile, the society's ongoing television production efforts, the formation of a nascent dot-com group to manage an early presence on the World Wide Web, a licensing program to brand lines of furniture, apparel, and outdoor gear, and other attempts to adjust to the new millennium contributed to what seemed to be a declining regard for what I came to call the G-word—*geography*. Geography was perceived as old-fashioned, stuffy, and static. Maps felt increasingly like an archaic backwater. I recall a senior public affairs person saying, "Can't we call it something like *world-ography*?" In retrospect, this is understandable: The attitude was a relic of an era spanning the final decades of the 20th century, when the old descriptive geography was in decline and many universities were abolishing their geography departments. (Today, Dartmouth is the only one of the Ivy League universities to retain its geography department.)

But the perception among National Geographic executives that geography was dull was deeply ironic. An intersection of technologies was revolutionizing the discipline. I summarized some of those changes on a page titled "Revolutions in Mapping" in the seventh edition *Atlas of the World*, published in 1999. I suggested that the *Atlas* "reflect changes in technology that with dizzying rapidity have turned the quiet art of mapmaking into a powerful tool for scientific investigation and for managing human societies and natural resources. Among these changes: computers that store vast archives of map data and render lines with super-human precision, software programs that turn maps into analytical tools, satellite imagery that combines photographic beauty with cartographic precision, global electronic networks that enable maps to stream across our ever-shrinking globe."

This publication was a landmark of sorts: We had teamed up with Esri

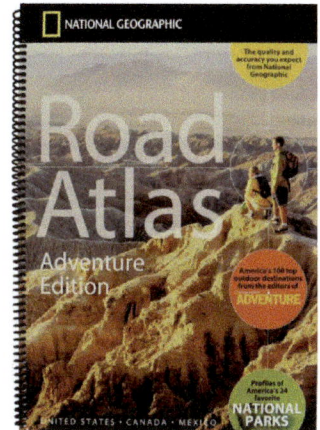

The *National Geographic Road Atlas*.

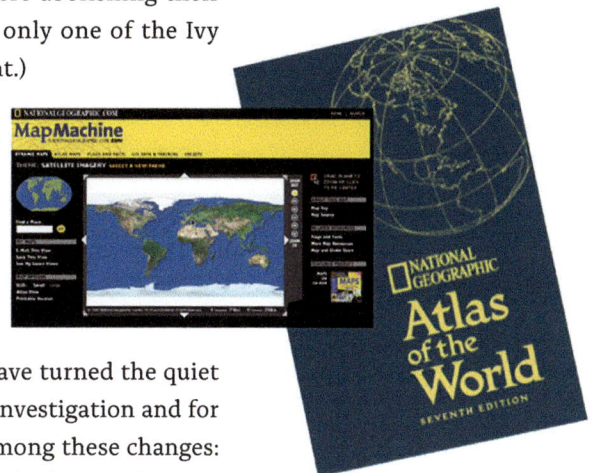

The seventh edition *Atlas of the World* and its online companion, MapMachine.

to produce **MapMachine**, an online interactive atlas as a digital companion to the book. We announced the *Atlas* publication and MapMachine launch simultaneously at an event at society headquarters. The MapMachine used Esri's nascent web-mapping technology to present multiple thematic maps within a searchable, pan-and-zoom interface. The map window was small, and its performance was sluggish by today's standards, but it was a big deal at the time. The Associated Press picked up the story and distributed it to news outlets. That, in turn, created a bit of a crisis on Esri's campus in Redlands, California. Web servers were briefly overwhelmed as thousands of people checked out the new resource. Unfortunately, MapMachine was ultimately retired, a victim in part of the rise of Google Maps and Google Earth, and perhaps as well from the G-word syndrome.

I'm happy to report that the G-word has long since shed its stigma at the National Geographic Society, and that geography is once again held in high regard.

The early success of MapMachine was the result of a strengthening alliance between Esri and National Geographic. A couple of years earlier, Jack and Laura Dangermond, founders of Esri, had invited a delegation from National Geographic to visit their headquarters in Redlands. Because of my role as the society's senior map person, I became the designated liaison

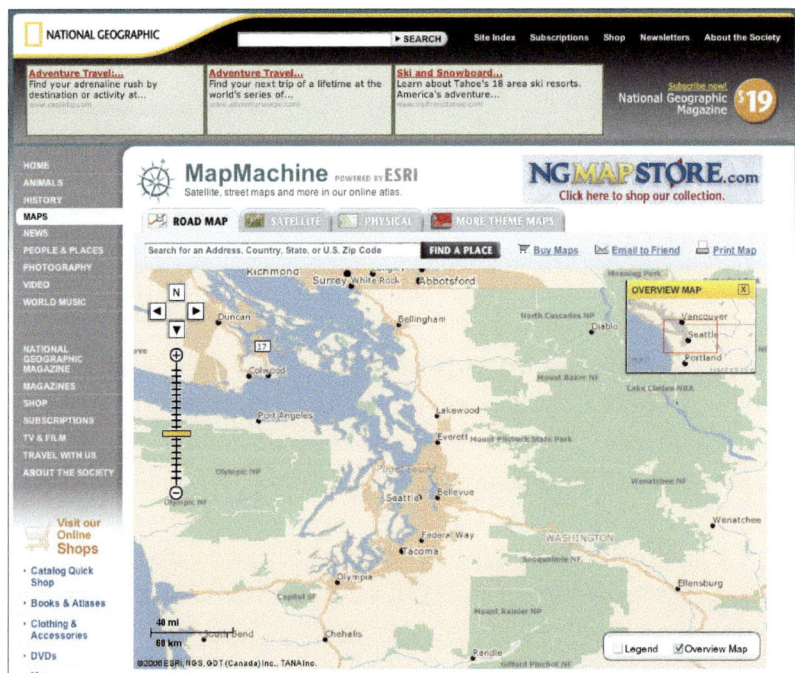

A later manifestation of MapMachine.

between the two organizations. It was a fortunate decision. I became good friends with Jack and Laura, for whom I have infinite affection and respect, and made several pilgrimages to their campus as we strategized about MapMachine and worked to enhance the GIS databases on which the *Atlas* was built.

The alliance with Esri also exposed me to the growing community of GIS professionals. I became a regular at Esri's User Conferences and was becoming increasingly aware of the spatial data that was being produced by thousands of GIS users across the globe. At this point, most organizations did their GIS work behind the scenes; sharing data online wasn't yet a common practice. I was (and still am) passionate about making people more aware of the wonders of the world, and it irked me that much of the data that organizations were creating wasn't being shared

WildWorld used World Wildlife Fund data to educate schoolchildren about Earth's biodiversity.

more widely. It was my opinion that data that was being held captive within organizations could be more broadly beneficial if it was "liberated"— that is, made available to broad audiences—and that part of the process of liberation should be explaining and interpreting the data or, in a sense, making data a good citizen.

I had a chance to indulge this offbeat passion through a partnership with World Wildlife Fund (WWF) and Ford Motor Company. I had encountered WWF data on world terrestrial ecoregions and immediately thought that it could be a means by which people would become more aware of the immense—and threatened—biological diversity of our planet. We produced a printed map and a website complete with photos and descriptions of about 900 ecoregions. And, thanks to Ford, we distributed 10 copies of a large-format, printed WildWorld map (*previous page*) to every K–12 school in the United States.

We continued to publish our traditional wall maps and experimented with some new directions. Our perennial best seller was our classic world map. It dawned on us that people might like a version of the map with a color palette better suited to den and office decors. The result: the Executive map with identical cartography but with a tilt toward earth tones. It became nearly as popular as the original (Classic).

## Toward a Nat Geo–branded storytelling platform

In 2006, I hired Frank Biasi as director of conservation and special projects. Frank had been director of worldwide conservation systems at The Nature Conservancy and combined technical knowledge with strong creative and entrepreneurial talents. We were looking for partnerships and projects that could bring in some revenue while aligning with the society's mission.

Frank joined us in part to support LandScope America, a collaboration with NatureServe, a nonprofit spinoff of The Nature Conservancy whose mission is to use science, data, and technology to guide biodiversity conservation. LandScope America was intended to support land conservation in the US with a rich mix of map resources and storytelling.

Frank and I also worked on two other ambitious concepts: the World Conservation Database and a *Story of the Earth* resource, both of which we envisioned as combining spatial data with interpretive multimedia storytelling. Although these grand visions did not achieve long-term success, they served to strengthen our interest in multimedia, place-based storytelling.

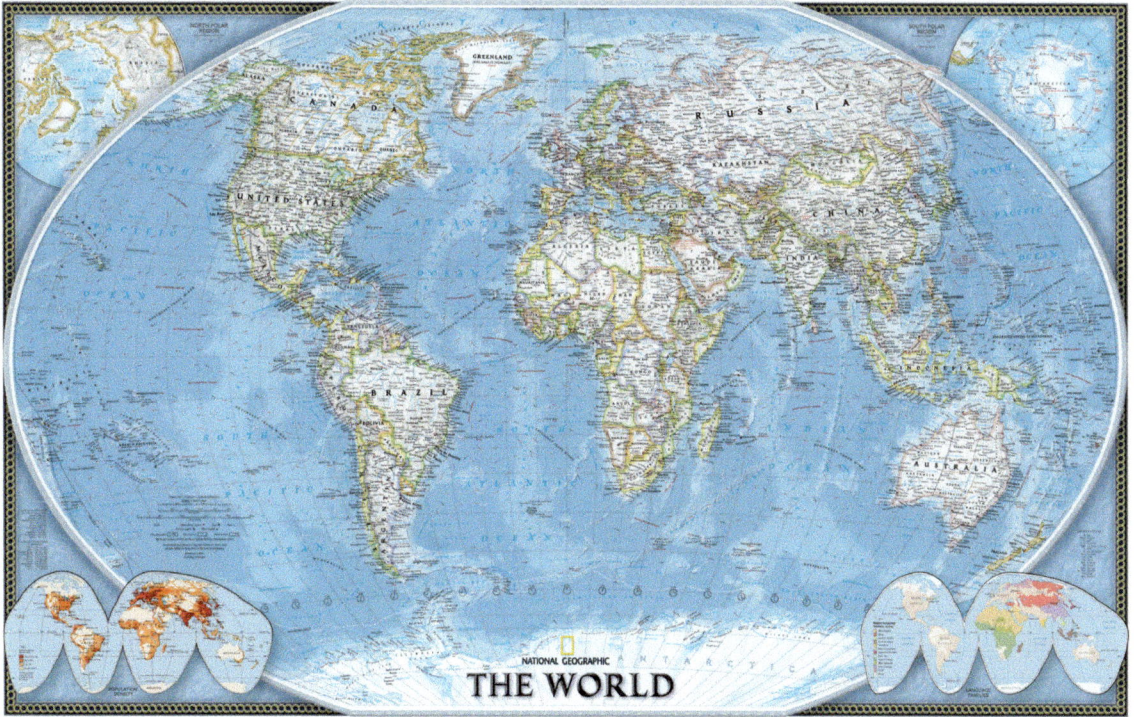

NATIONAL GEOGRAPHIC
# THE WORLD

NATIONAL GEOGRAPHIC
## THE WORLD

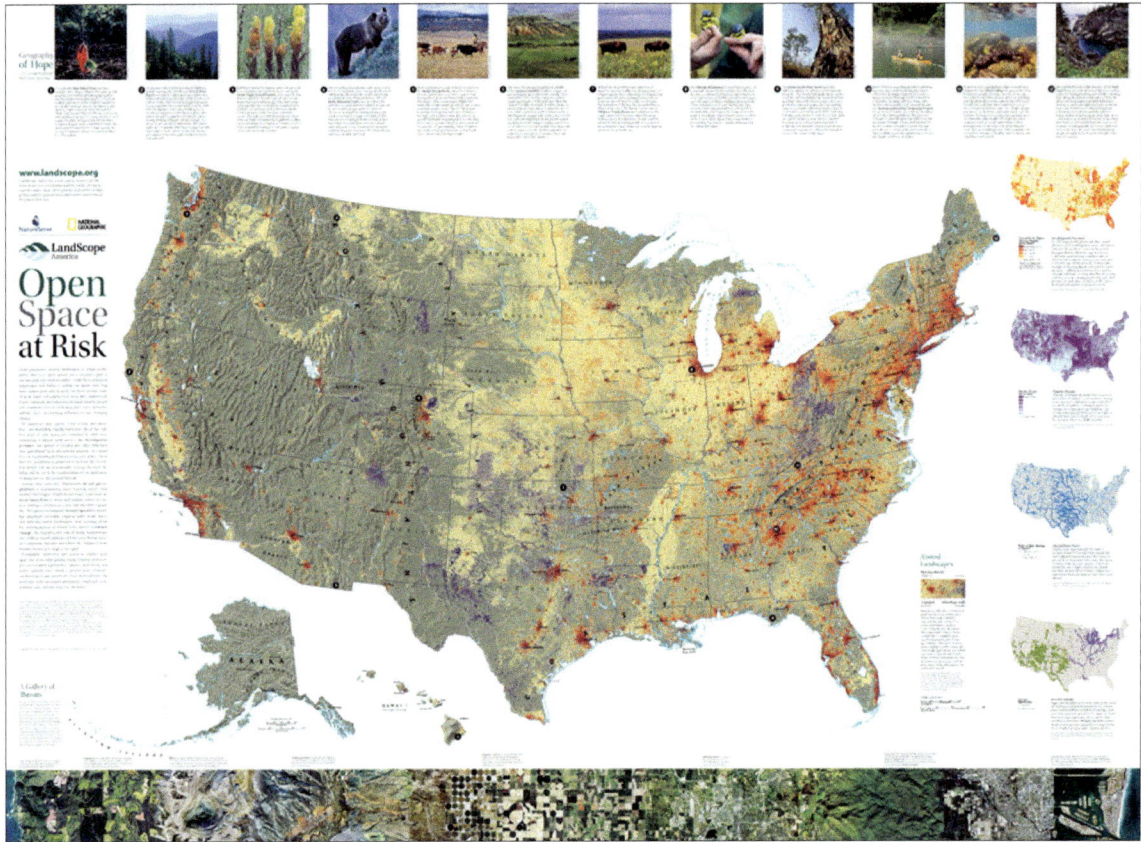

A printed map of the US conservation landscape was part of the LandScope America project.

These efforts led us to think about resources that we could develop for the web that were particularly appropriate to the National Geographic brand. It was obvious to us from the start that the society was renowned for two things—stunning images and beautiful maps—and that digital media could give us an opportunity to combine these assets to tell highly engaging interactive stories. The idea was to take readers on virtual expeditions, intuitively guiding them from one location to the next with mouse clicks and map interactions.

So we began developing a concept that we came to call **GeoStories**. We worked with Old Town Creative (now renamed Elebase), a small group in Montana that had developed a spatially enabled content management system (CMS). We designed a format that would allow authors and other creative people to add photos, text, and interactive maps to a database, the contents of which would be displayed in a three-panel viewer that, in turn, could be easily embedded in websites. It was our hope that GeoStories would be embraced by the society's dot-com team and that external partners would pay a subscription fee to use GeoStories on their own websites.

Prototype GeoStory, presenting place-based stories combining text, images, and maps.

GeoStories enjoyed modest but significant early success. The society's news team published a battery of stories, and a few of *National Geographic Magazine's* local-language partners embraced them for their Nat Geo–branded websites. GeoStories were licensed to several external organizations, including REI.

The society quietly retired GeoStories several years later. But the lessons I learned about web-based, map-enabled storytelling proved to be enormously valuable in the next chapter of my career. I consider GeoStories to be a predecessor to the multimedia storytelling tools that we developed—and that have found considerable success—at Esri.

I continued to work closely with Esri throughout my final months at National Geographic, sharing Jack Dangermond's conviction that our organizations were deeply aligned despite profound differences in corporate culture and outlook.

## From National Geographic to Esri

It was becoming clear to me that I had more kindred spirits at Esri than at National Geographic. Jack had kindly indicated that I was welcome to join his company, and in fall 2010 I made the jump. I'm proud to have worked at the National Geographic Society for 27 years, and despite feeling some regret for leaving the society, I knew that Esri, despite being a for-profit company, was driven by a deep passion for harnessing and extending the

Jack and Laura Dangermond with National Geograpic Society chairman Gil Grosvenor at the 2010 Esri User Conference. Gil worked diligently for decades to boost geography education in K–12 schools.

power of geography to help ensure a sustainable future. I could continue to indulge my starry-eyed idealism, but in a stable, supportive—and profitable—environment.

I moved my various atlases and wall maps to Esri's regional office in the Washington, DC, suburbs in October 2010. (Jack had urged me to move to California but family ties dictated otherwise.) I was to report to Clint Brown, director of software products, and my mandate included assisting with Esri's web resource, ArcGIS Online, and developing a vague concept that had been called "map stories" or "map of the month." Clint helped me recruit a small team, and I was soon working with Lee Bock, an Esri veteran and talented developer, and David Asbury, a GIS whiz who had moved to DC after working for a small salmon conservation nonprofit in California. Not long thereafter, we brought on Steve Sylvia, a recent graduate of Penn State's geography program, as a summer intern. We were looking for a web developer, and at one point Steve, who had almost no experience doing development work, volunteered to learn some skills. Almost overnight, he became a self-taught programming whiz. A dozen years later, Steve, David, and Lee are all still members of the team and doing excellent work on its behalf.

We began to experiment. We would assign ourselves a topic, sketch some design concepts, and assemble appropriate maps, text, and images. Most important, we'd create a user experience that was tailored specifically to the topic we had chosen. It might not have been the most efficient way to get our storytelling efforts off the ground—Jack grew impatient with us after several months of experimentation—but it grounded us and inspired some out-of-the-box thinking.

One of the earliest and most rewarding of our many collaborations with nonprofit organizations was a joint effort with the National Audubon Society called **Beating the Odds: A Year in the Life of a Piping Plover** ⊖. Only about 8,000 of these diminutive shorebirds are left, due in part to the fact that they nest on the ground in simple "scrapes" near beach grasses, making them vulnerable to predators, sun bathers, pets, and vehicles.

The story was a chance for me to indulge my passion for birds and conservation and to support an organization I had long admired. In fact, years earlier I had been an Audubon employee thanks to a summer job—earning $50 a week—at the society's Corkscrew Swamp Sanctuary, which protects a breathtaking old-growth cypress swamp in southwest Florida. I performed odd jobs, the best of which was to do early-morning circuits through the swamp to sweep otter poop off the boardwalk. I saw a lot of otters,

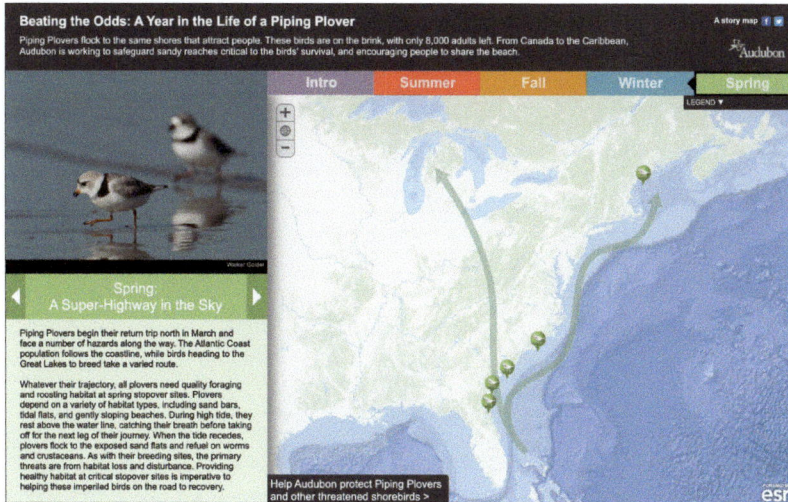

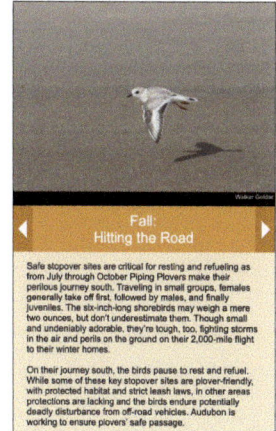

alligators, snakes—and wonderful birds. The following year my meager earnings helped fund my western adventure (see "Travels With Godzilla" on page 88).

In creating the story, our little team learned a lot about integrating maps and multimedia. It was perhaps the first time that we had made maps pan and zoom in sync with the narrative. It's looking a little dated now, but it was quite effective in depicting the hazardous wanderings of a special creature.

The second experiment was **Geography, Class, and Fate: Passengers on the Titanic** ⬅→ (*next page*). We became aware in the spring of 2012 that the 100th anniversary of the sinking of the *Titanic* was just a couple of weeks away. After some head-scratching about how we could commemorate the event, we found the ship's entire passenger manifest on Wikipedia, complete with information on points of origin, intended destinations, and class. So we mapped the data, and to our amazement, we discovered some unexpected patterns. Passengers in first class were primarily coming from, and heading to, major cities in western Europe and the United States. And their survival rate was relatively high: Almost 200 of the 324 first-class passengers survived. Steerage passengers, however, came from scores of villages in the British Isles, Scandinavia, the Balkans, and the Levant. And barely a quarter of them survived the sinking. A popular blog picked up the story; that, and the story's timeliness, were factors in it becoming the first of our efforts to find a large audience.

A third experiment was the **Twister Dashboard: Exploring Four Decades of Violent Storms** ⬅→. The National Oceanic and Atmospheric Administration

*Beating the Odds: A Year in the Life of a Piping Plover* portrays the seasonal movements of a threatened species.

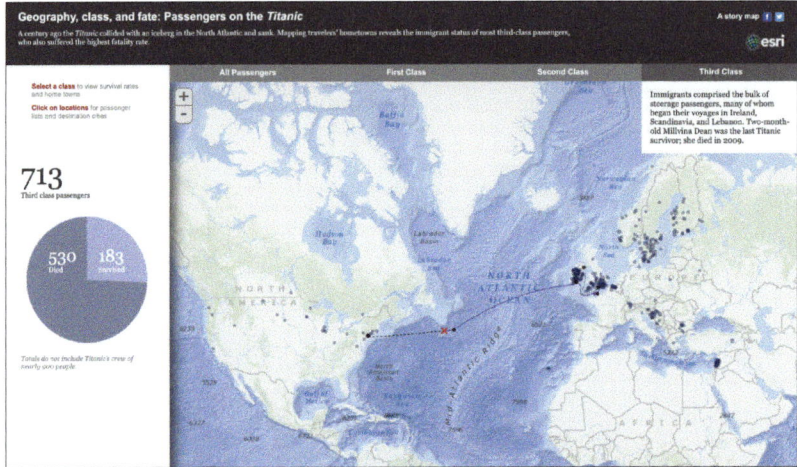

*Geography, Class, and Fate: Passengers on the Titanic* reveals stark differences between steerage (*top*) and first class (*bottom*).

*Twister Dashboard: Exploring Four Decades of Violent Storms* lets readers explore a database of hundreds of tornadoes.

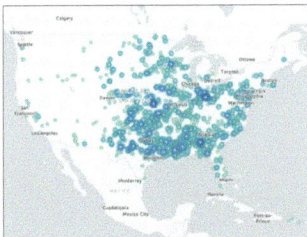

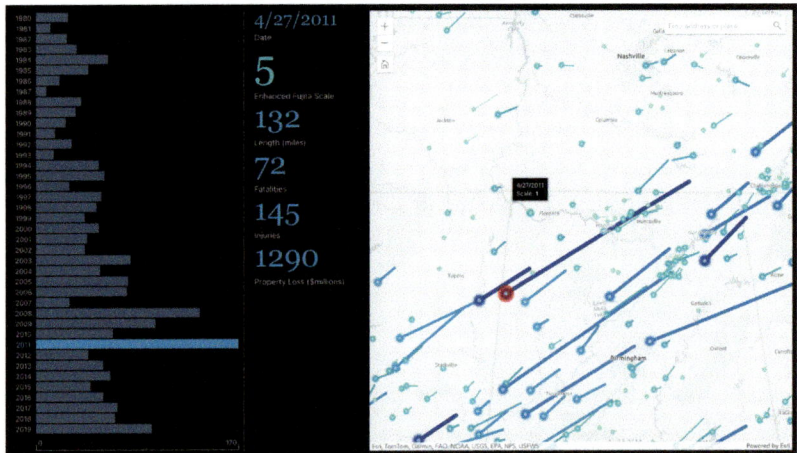

(NOAA) keeps a public database of all US tornadoes, including date, severity, number of fatalities, and track. We came up with a design for an interactive dashboard (*above*), and Lee executed it elegantly. A bar graph of tornadoes by year updates dynamically as readers pan and zoom on the map; clicking on an individual tornado reveals relevant statistics.

These and other efforts helped us navigate a learning curve, and although some of our experiments were more successful than others, we came up with several clever and effective solutions. We've retired some of our early efforts because they no longer work on modern browsers, but the early projects described above are still active and are as relevant today as ever. Although most of the user experiences we designed worked reasonably well, they solved one-off problems and visualized data that was structured in a particular way. They didn't, however, enable others to easily pour their own data and multimedia content into them.

That began to change in 2012 with another early prototype. I took a trip to New York City, where I walked and photographed the High Line, a popular linear park built atop an abandoned elevated train structure. We created a format that included an image window with a text panel, a map, and a carousel of thumbnails. Readers could navigate by clicking map icons, thumbnails, or previous and next pointers. **A Walk on the High Line** ⊖ debuted the Map Tour format.

We developed other web apps during this period that provided simple functionalities. Swipe and Spyglass enabled users to explore two related maps. We created a Side Panel app, in which authors could feature a single map and interpret it in the accompanying panel. These apps weren't sophisticated enough to enable narratives, which meant that they weren't, strictly speaking, storytelling solutions. But we subsequently developed Playlist and Countdown apps that offered a numbered series of map-and-side-panel combinations.

The GIS community was starting to become aware of our efforts. I remember demonstrating the Spyglass app with Esri veteran Bern Szukalski at the 2012 Esri User Conference and eliciting audible gasps from the audience.

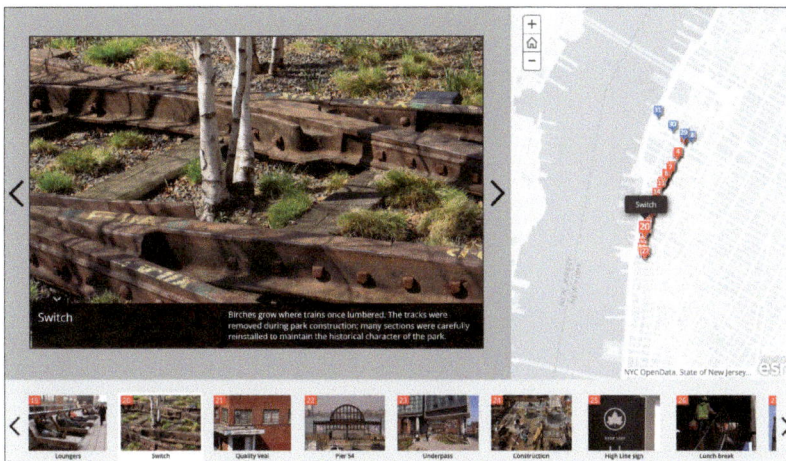

Map tour

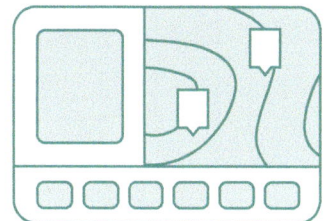

Map tour of the High Line, New York's aerial park.

## Swipe

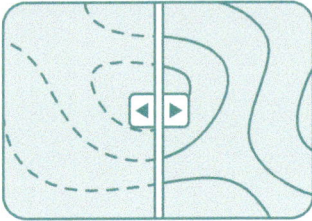

Swipe showing Washington, DC, in the 19th century and today.

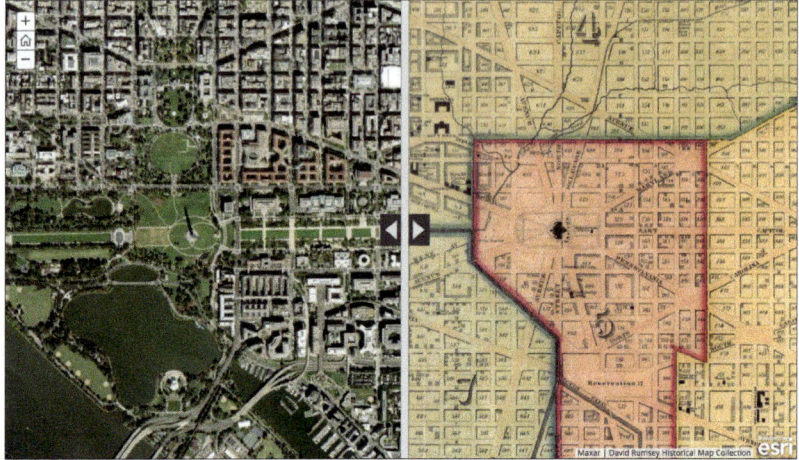

## Spyglass

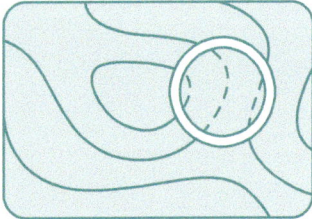

Peering through modern Manhattan with a spyglass to glimpse a 19th-century map.

## Playlist

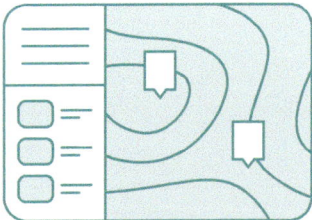

Playlist story on the most-visited national parks.

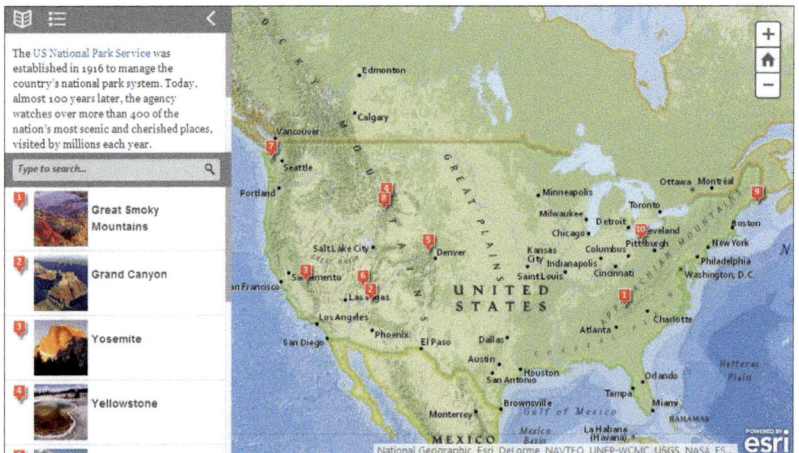

## Facilitating storytelling: Builder functions

It was at about this time that a young French developer joined our team. Gregory L'Azou was to have a profound impact on our early success. He convinced us that we could enrich our apps with builder functions that would allow authors to create stories using an intuitive, WYSIWYG (what you see is what you get) user experience. No downloading, no web development skills required. Previously, creating a story required multiple steps: developing narrative content and importing it into a specific format (usually a spreadsheet), downloading and configuring web files, and deploying the app to a hosting environment. The last of these steps was the biggest hurdle for authors who didn't have access to a web server or experience in administering a website.

Our **Map Tour** and its accompanying builder function debuted at the 2013 User Conference. It's no coincidence that the popularity of multimedia stories started to take off at this point. In 2012, we had counted 120 stories hosted on Esri's cloud service, ArcGIS Online. In 2013, the number reached 1,200. In 2014, it was 12,000. We had no idea that, a decade later, hosted stories would number in the millions. But we were starting to gain momentum.

The next breakthrough was **Story Map Series**, which we developed in 2013 and 2014. We ultimately created three versions of this app to provide navigation options: **tabbed**, **bulleted**, and **side accordion**. As with our other apps, we prototyped and tested this new format by producing our own stories. In all the versions, the side panel contained the narrative text but could also accommodate images and graphics. We anticipated that Series would be useful for presenting a sequence of related maps. That's exactly how many of our users employed it; however, Series became quite popular as a way of making a collection of maps easily accessible using a single web address. For example, firefighters and other emergency responders could quickly access dynamic maps showing wildfire extents, aerial imagery, and other frequently updated resources.

The next year brought **Story Map Journal**, our most sophisticated and seamless narrative format yet. Journal presented text and other content in a continuously scrolling side panel; the remainder of the screen, or *main stage*, would refresh with each section of the story to display maps, images, and video. Journal proved highly popular with our growing community of storytellers.

Meanwhile, our team was gaining new members, among them a delightfully eccentric British expatriate named Rupert Essinger. He became one

## Series: tabbed

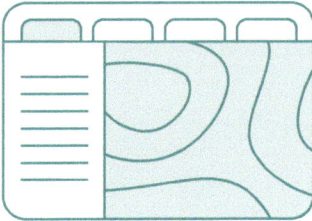

Tabbed story examining the aftermath of Hurricane Katrina.

## Series: bulleted

Bulleted story showing the most damaging hurricanes in US history.

## Series: accordion

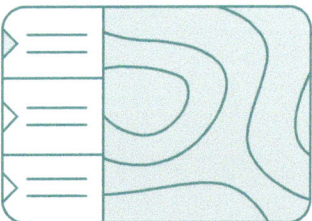

Accordion story looking at renewable energy world-wide.

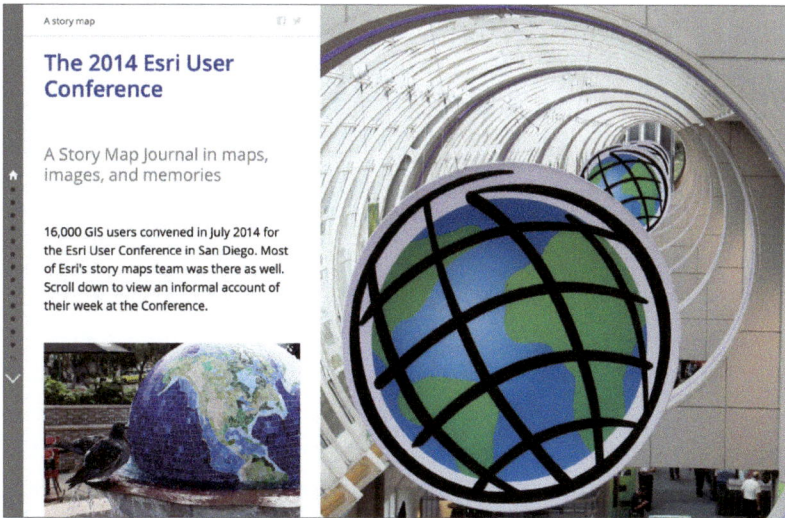

Journal

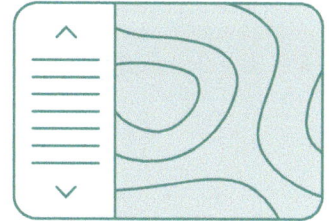

Story Map Journal account of the 2014 Esri User Conference.

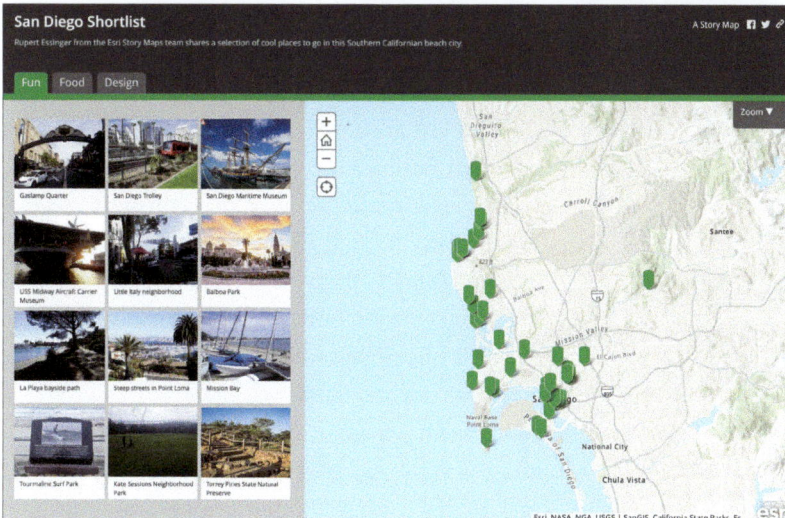

Shortlist

Story Map Shortlist: Rupert Essinger's initial concept (*below*) and final execution (*left*).

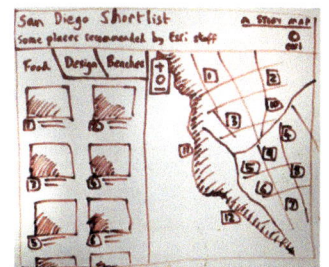

of our most passionate advocates, gaining a reputation for attracting audiences, Pied Piper style, to his impromptu demos at Esri User Conferences. He presented us with a concept for a form of nonlinear storytelling, which he called Shortlist. The idea was to map and summarize points of interest in and around a destination and to categorize them into themes that users could access using tabs. As we had hoped, we began to see Shortlist stories of multiple cities and other attractions, ranging from parks and game reserves in Kenya to historic cemeteries in Rutherford County, Tennessee.

Greg L'Azou was excited about our progress—especially the Journal app—but he immediately began lobbying for more. He wanted a format in which

Cascade

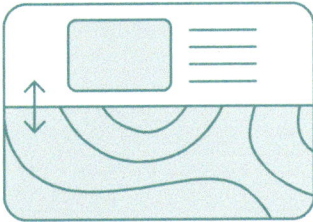

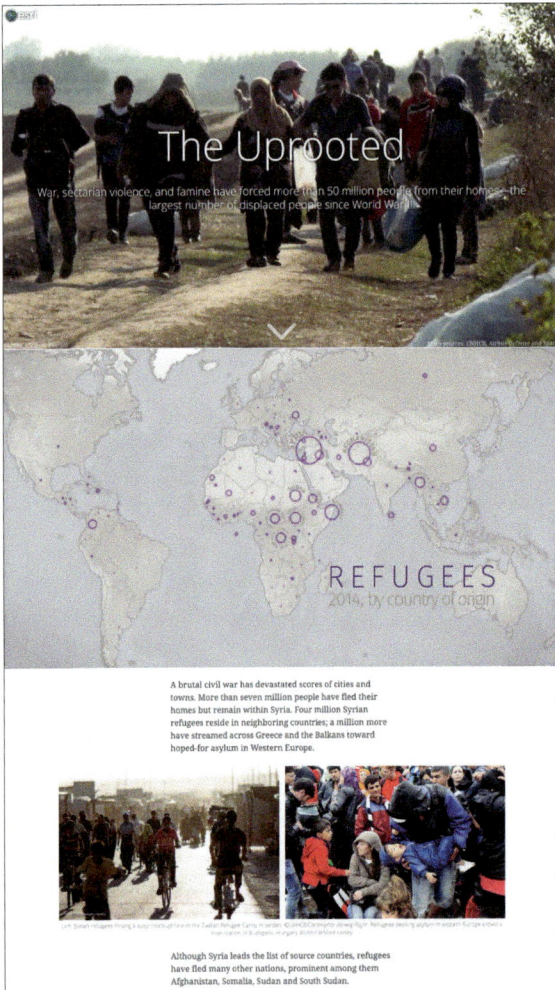

Debut of the Story Map Cascade app: tracing refugee migration across Europe.

the whole browser window scrolled, not just the side panel. I had trouble picturing this at first, but we were soon sketching concepts and creating mockups. Central to this new app would be the ability to insert immersive sections that would behave much like the Journal app, with side (or floating) panel scrolling and the remainder of the screen presenting maps and images in sequence. We released the new format as **Story Map Cascade**.

We decided to do a story that would test the limits of the Cascade app and show off its features. Europe and the Middle East were in a refugee crisis at the time, so we decided to pull out all the stops on a story that mapped the exodus of hundreds of thousands of refugees while also putting a human face on the crisis. The result: **The Uprooted** ⮌, which opens with a slow-motion, looping video that sets the mood for a narrative that included scrolling text and images, full-width static maps of refugee populations, and a 3D immersive section that followed the main migration path from the Middle East to northern Europe.

I'm particularly proud of this story for its combination of factual information and emotional impact—which to me epitomizes the utility of place-based multimedia storytelling to explain complex phenomena while engaging our emotions and spurring our empathy.

With the release of Cascade in 2016, we had an array of apps that gave our users a variety of ways to combine maps and other content, ranging from a simple map viewer to a sophisticated tool for multimedia storytelling. By this time, more than 100,000 stories were hosted in ArcGIS Online, and the numbers were growing rapidly. Our presentations and workshops were often standing room only. Storytelling apps were being enthusiastically embraced by the GIS community. They were also being discovered by educators and were being used for instructional purposes, as well as for student projects, providing an alternative to traditional—and static—research papers.

# The arrival of ArcGIS StoryMaps

That was the bright side. On the other hand, our array of storytelling apps was confusing to some people. Storytellers would start to create a narrative using one app, only to realize that a different app was a more effective solution, requiring them to start over. Because we had learned a lot as we sequentially developed our apps, their builder functions were inconsistent from one app to another. Stories performed well on large screens, but the experience of viewing them on mobile devices was often unrewarding. Finally, the oldest of our products—especially Map Tour—were starting to look dated.

We were close to becoming victims of our success. I was pretty good at developing new ideas and creating stories, but my skill at managing the rigors of software development and maintenance was rudimentary. As our audience expanded, and as our team grew, we reached a crossroads that mandated two changes in strategic direction. First, the Story Maps team became two sister teams: An editorial group, which I would lead, would promote story maps and nurture a burgeoning storytelling community with workshops, webinars, presentations, exemplary stories, blog posts, and more. A sister development team would maintain and enhance the storytelling resources themselves. The two teams would coordinate closely; the editorial team, for instance, would advocate for enhancements based on its close association with the storytelling community; the dev team would continually enhance and refine the product. As ArcGIS StoryMaps evolved, the editorial team would hold workshops and produce instructional materials to socialize recent enhancements.

Sathya Prasad took over management of the sister group. Sathya, a brilliant software researcher, developer, architect, and engineering manager, began his career at Esri in its Applications Prototype Lab.

The second strategic shift was to work toward a single, second-generation product, ArcGIS StoryMaps, that would ultimately combine nearly all the capabilities of what would now be called "classic" storytelling apps—Series, Journal, Cascade, and so on—into a single product.

The concept for ArcGIS StoryMaps was to organize its various media components into discrete items called blocks; authors could access these blocks through a builder menu called the block palette. Storytellers could assemble their narratives by arranging a sequence of blocks representing basic items: text, buttons, separators; media items, including images, video, and embeds; and immersive items. The latter included sidecar and map

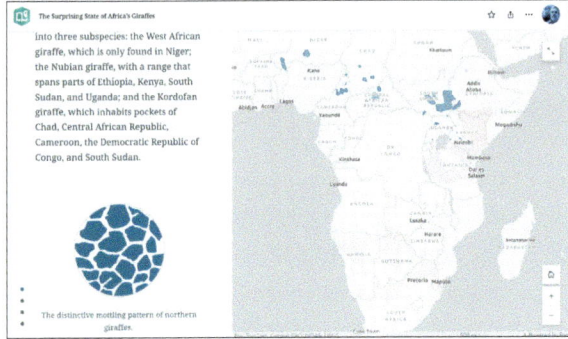

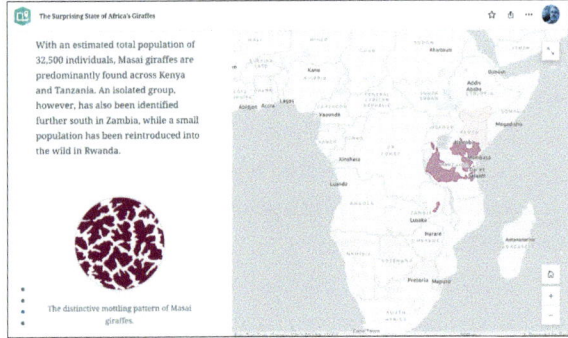

ArcGIS StoryMaps arrives with a team-produced story on giraffes.

tour formats that were descendants of the more immersive functions of the classic apps.

One of the media categories was maps. ArcGIS StoryMaps featured two mapping capabilities. One enabled authors to take advantage of the rapidly expanding content and functions of ArcGIS Online and its map viewer; authors could assemble and refine multilayered web maps to incorporate into their stories, using the builder to set map scales and extents and determine which layers would be made visible. The other was called **express maps**, which had begun as an independent project, code-named Mercator, to create a simplified mapping function expressly designed to serve users with limited or no experience in GIS. Intuitive features enabled authors to create simple maps incorporating points, lines, areas, labels, arrows, and other items.

ArcGIS StoryMaps became available in beta on April 8, 2019. It went into general release on July 2 of that year, just before the Esri User Conference, the temporal fulcrum around which all of Esri swivels. The functions of this new, second-generation storytelling tool were somewhat limited at first, but Sathya's team had embraced a policy of providing frequent updates, with new releases occurring roughly every two weeks and automatically incorporated into all Esri-hosted stories. Over the succeeding

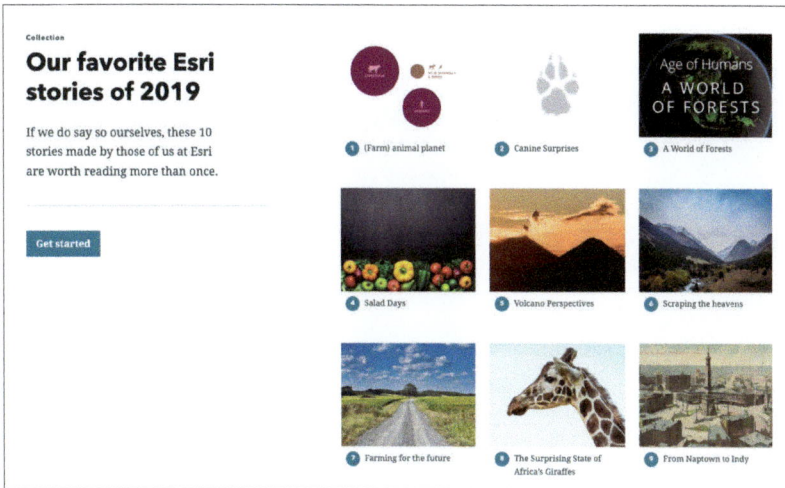

Collection

A collection page aggregates a selection of Esri-produced stories.

months, ArcGIS StoryMaps became continually more versatile as it incorporated more of the features that had formerly been distributed across an array of apps—as well as new features unique to the second-generation platform.

Among the first stories our team created using ArcGIS StoryMaps was **The Surprising State of Africa's Giraffes** ⊖. Its sidecar section was an ideal format for displaying range maps and marking patterns of the various species and subspecies of giraffe.

Among the more significant enhancements to ArcGIS StoryMaps was the addition of collections, simple gallery pages that aggregate up to 60 individual multimedia stories. Collections offered three layout options and had an intuitive builder function. Story authors used collections to display stories on related topics or produced by a single organization. Collections could also be used as a table of contents, with individual stories serving as chapters in a longer narrative.

A more recent enhancement of collections was the option to add an index map with clickable location icons to any of the layout options, providing an alternative means of access to items with location information.

ArcGIS StoryMaps also enables authors to customize the look and feel of stories with **themes**. Initially, authors could choose from a menu of six options, each with its own combination of colors, fonts, and other design features. A single click would change the story's appearance. Later enhancements gave authors an almost unlimited selection of options using a theme builder that provided access to hundreds of fonts and thousands of colors, as well as options for buttons, separators, and other elements.

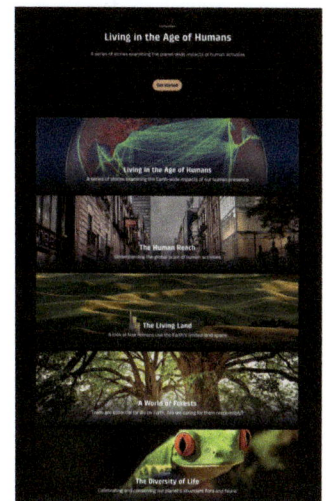

A collection of team-produced stories examining the profound impacts of human activities on the planet.

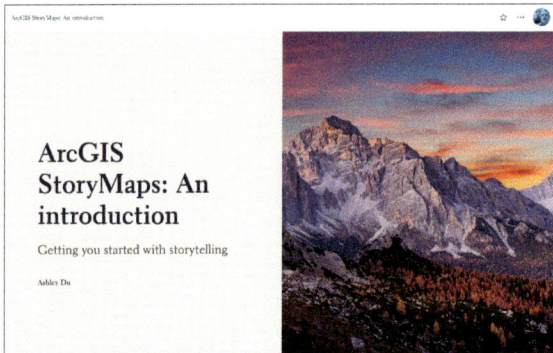

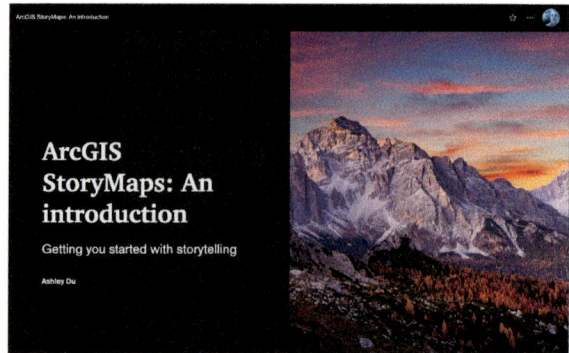

Themes offered single-click styling of stories and the ability to explore almost unlimited options for visual styles.

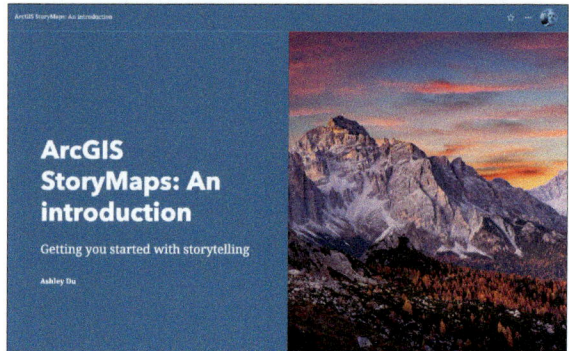

Authors could save and share themes, facilitating branding and visual consistency across multiple stories and collections.

An additional output type arrived in 2024: **briefings**. In contrast to the conventional ArcGIS StoryMaps scrolling experience, briefings are a slide-driven format designed to facilitate short, in-person presentations. Briefings are integrated with a companion tablet app that enables users to create a duplicate of their presentation that can be shown in the field—even without a web connection.

While stories are designed for readers to immerse themselves in narratives, briefings focus on a familiar, slide-based format to present information in concise text, bullet points, and numbered lists. Briefings are an ideal way to share concise, impactful information. Like ArcGIS StoryMaps stories, briefings accommodate interactive maps and 3D scenes as well as embeds, videos, and images. Authors can add dynamic content from multiple sources directly to their slides or in an attachment link that opens within the presentation. A variety of slide layouts gives authors multiple options for combining text and multimedia. Presenters can also style their briefings using the same theme builder that supports stories and collections.

Cover slide (*left*) of an ArcGIS StoryMaps briefing. Above, viewing the briefing on its tablet app.

Another relatively recent arrival is templates, which allow authors to create skeletal stories that others can access, duplicate, and use to author narratives with a structure and graphic style that reflect the template's characteristics. Key to this function are instruction blocks, which guide formatting and suggest media elements, giving template users a head start in their storytelling process. Instruction blocks appear within tinted boxes and disappear when templated stories are published.

Templates can serve multiple purposes. Consider this example. Let's say I'm a manager in municipal government. I might want my departments to keep the public informed by regularly publishing summaries of capital improvement projects and other initiatives in the form of multimedia stories with consistent branding and structure. A projects template makes it simpler for my teams to create and publish reports. They won't need to learn all the builder functions in ArcGIS StoryMaps to create their reports (although those additional features will remain available). Rather, they can cut and paste text, alter locator maps, and upload images as directed by my story template.

If I'm a teacher, I can create a school report template that's specific to the curriculum items I'm currently presenting. I can customize formats, headings, and immersive sections to suit my needs and then make the template available to my students to create their reports. It will be easier for me to assess my students' work, because their reports will all align with the structure I created when I built my template.

Over the years our users have frequently asked for ways to refine and customize stories beyond the considerable flexibility that ArcGIS StoryMaps

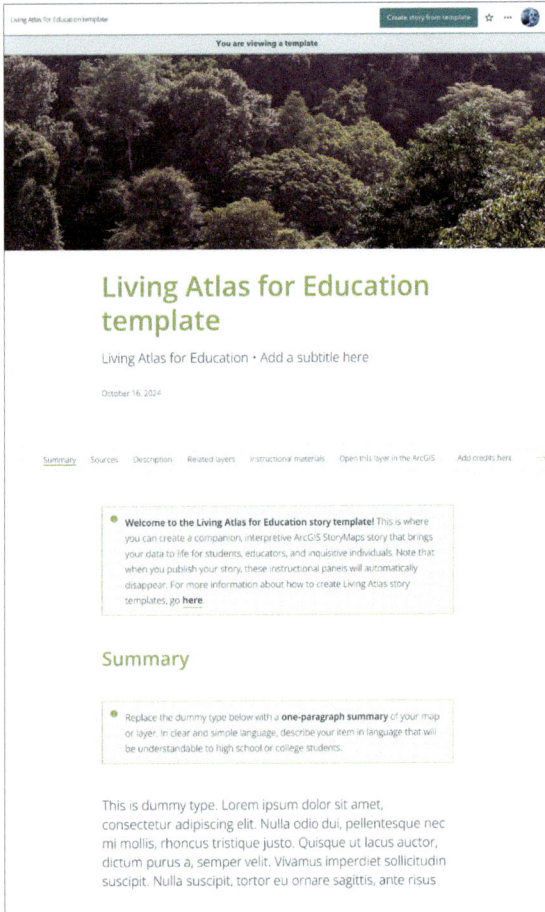

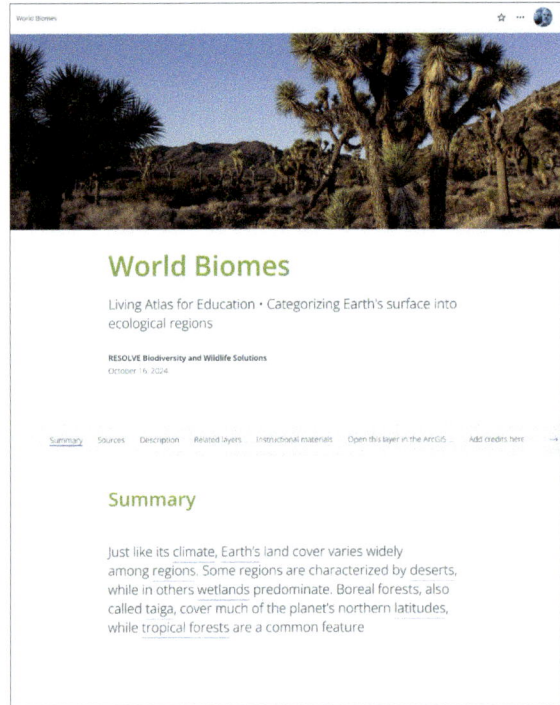

A template (*left*) presents a story structure and offers instructions for creation of stories. Instruction blocks disappear when the template-based stories are published (*right*).

offers by default. Some, for example, have sought to create custom URLs to replace the standard storymaps.arcgis.com/stories/[name] web address; others wanted to swap in their own type fonts and special logo treatments.

After a good bit of technical head-scratching, our developers came up with an ingenious solution that we call **advanced embedding**. With just a few lines of code that Esri provides, you can embed an ArcGIS StoryMaps story directly in your own web page. The web page serves as a "wrapper" for your story. You can then make all sorts of modifications, including customizing headers and footers, altering formats within the story (column widths, typographic refinements, and the like), and using analytics services other than the currently supported Google and Adobe services. The story itself resides in ArcGIS Online, so it will benefit from frequent product enhancements and updates.

Advanced embedding amounts to a breakthrough, making it possible for stories created with ArcGIS StoryMaps to appear fully and seamlessly integrated with organization websites in both appearance and function.

Enhancements such as advanced embedding might be game changers, but other, more modest, enhancements can have big impacts. Two examples: the ability to fine-tune type sizes within stories and an option to change the spacing between the blocks, or elements, of a story. These relatively simple functions can have a dramatic effect on the look and feel of your stories.

The journey of ArcGIS StoryMaps continues. The storytelling community is steadily growing; when this book went to press, storytellers were creating an average of 4,000 stories per day. Meanwhile, our team was adding items to its list of proposed new features at much the same rate at which it releases new functions. As technology evolves and the needs of our users change, ArcGIS StoryMaps will continue to grow, serving a diverse and international user community with ever more powerful storytelling tools.

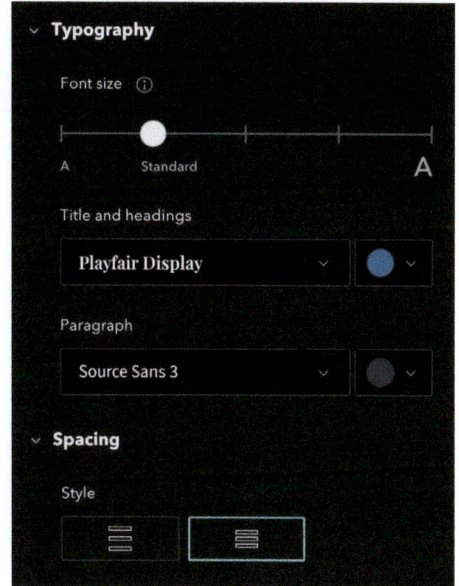

Typography and spacing controls within the ArcGIS StoryMaps theme builder.

# Masters of Tradition

A cultural journey across America

**Author:** Smithsonian Center for Folklife and Cultural Heritage

**Medium:** Custom multimedia story

**Story behind the story:** Marjorie Hunt, a curator at the Smithsonian Center for Folklife and Cultural Heritage, collaborated with the National Endowment for the Arts (NEA) to feature scores of traditional artists, craftspeople, and performers who had been honored by NEA as National Heritage Fellows. The ArcGIS StoryMaps editorial team helped design and build the resource.

**Why it's special:** The more than 150 people profiled in this multimedia narrative illustrate the diversity of cultures and backgrounds that are a vivid reminder that our nation is strengthened and enriched by its cultural diversity. Individual profiles include text, images, video, and audio.

The site was honored twice by the American Association of School Librarians as one of the best digital tools for teaching and learning.

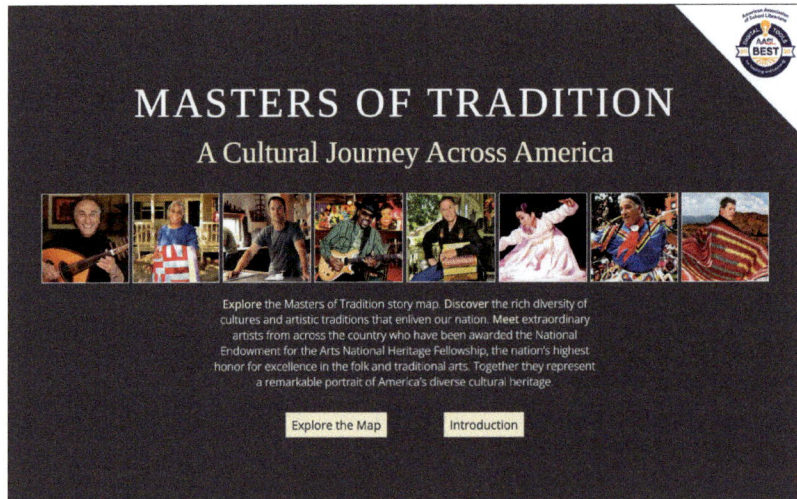

MASTERS OF TRADITION
A Cultural Journey Across America

Explore the Masters of Tradition story map. Discover the rich diversity of cultures and artistic traditions that enliven our nation. Meet extraordinary artists from across the country who have been awarded the National Endowment for the Arts National Heritage Fellowship, the nation's highest honor for excellence in the folk and traditional arts. Together they represent a remarkable portrait of America's diverse cultural heritage.

Explore the Map          Introduction

*Top*: A landing page links to the map and an introductory essay.

*Clockwise from left*: Giuseppe and Raffaela De Franco, Italian-American musicians and dancers; Vanessa Paukeigope Jennings, Kiowa regalia maker; Canray Fontenot, Creole fiddler. Photos by Tom Pich.

An interactive map and summary list (*bottom*) provides access to Heritage Fellow profiles. Each featured Heritage Fellow has a biographical article that begins with a portrait photo, such as Mary Lee Bendolph, African-American quilter (*below*). Readers scroll down to find a descriptive essay and supporting photos, videos, and audio for each individual (*right*). Photos by Tom Pich.

## Mary Lee Bendolph                                   ✕

**African American Quilter**
2015 NEA National Heritage Fellow
Boykin, Alabama

Quilters of Gee's Bend, Alabama: (left to right) Loretta Pettway, Lucy Mingo, and Mary Lee Bendolph. Photo by Tom Pich.

"Strips and Strings" quilt made by Mary Lee Bendolph. Photo courtesy of Souls Grown Deep Foundation, Steve Pitkin/Pitkin Studio

Masters of Tradition: A Cultural Journey Across America

Search
*by name, place, tradition, culture group, year*

**Sheila Kay Adams**
Appalachian Ballad Singer, Musician, and Storyteller
2013 | Marshall, North Carolina

**Obo Addy ***
Ghanaian American Drummer and Singer
1996 | Portland, Oregon

**Rahim AlHaj**
Iraqi American Oud Player and Composer
2015 | Albuquerque, New Mexico

**Juan Alindato ***
Puerto Rican Carnival Mask Maker
1987 | Ponce, Puerto Rico

**Michael Alpert**
Yiddish Musician and Tradition Bearer
2015 | New York, New York

**Anjani Ambegaokar**
Indian American Kathak Dancer
2004 | Diamond Bar, California

*Deceased

Leaflet | Powered by Esri | National Geographic, DeLorme, HERE, UNEP-WCMC, USGS, NASA, ESA, METI, NRCAN, GEBCO, NOAA, increment P Corp.

# City of Irvine Great Park

## Framework plan

**Author:** City of Irvine, California

**Medium:** ArcGIS StoryMaps

**Story behind the story:** The City of Irvine, California, has grand plans for a Great Park that uses land made available by the 1999 decommissioning of the Marine Corps Air Station El Toro.

**Why it's special:** "Framed by forest, defined by open space and a wide variety of community-serving uses," the city announced, "it brings nature, exploration, and wonder to Irvine's doorstep ... The Great Park is a celebration of both Irvine's rich heritage and diverse community character—and a multifaceted investment in our collective future."

Irvine used the ArcGIS StoryMaps map tour immersive experience to take its citizens on a virtual stroll through the 1,300-acre park, which is opening in phases. The tour is presented atop a map that superimposes a colorful rendering of the park plans on an imagery base.

Hundreds of towns and cities present their city plans, improvement projects, and amenities as multimedia narratives; this one sets itself apart from other municipal stories mainly in the ambitious scope of Irvine's dreams.

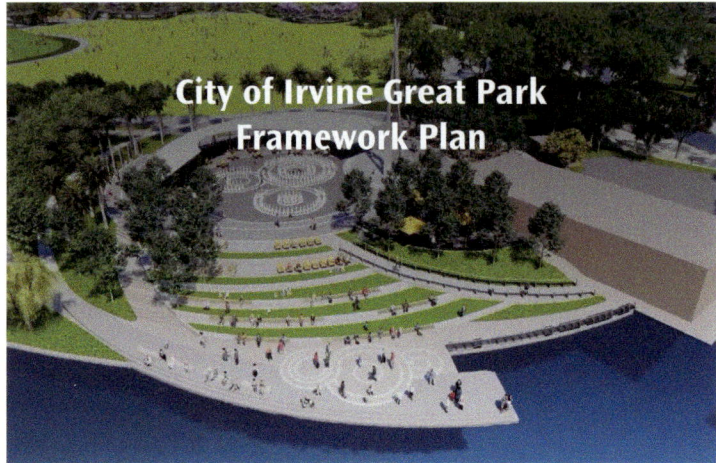

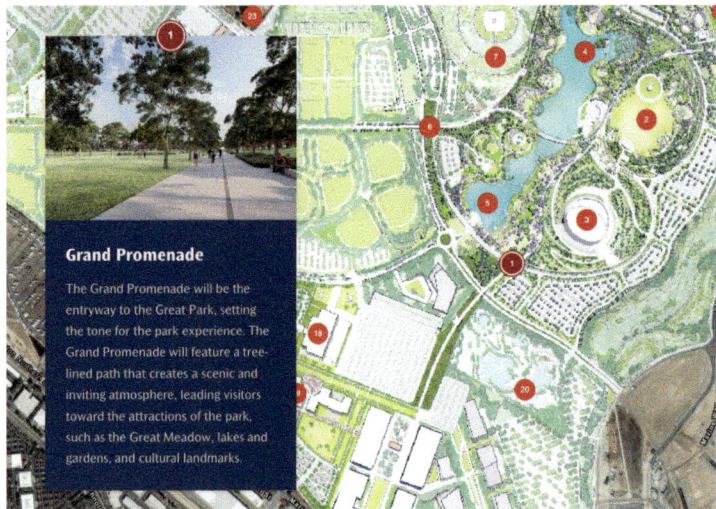

**City of Irvine Great Park Framework Plan**

**Grand Promenade**

The Grand Promenade will be the entryway to the Great Park, setting the tone for the park experience. The Grand Promenade will feature a tree-lined path that creates a scenic and inviting atmosphere, leading visitors toward the attractions of the park, such as the Great Meadow, lakes and gardens, and cultural landmarks.

The story's title panel (*top*) features a looping video of the park's major landmarks. The map tour (*above*) presents conceptual drawings and interpretive text within floating panels.

*Facing page*: A selection of park features presented in the Great Park map tour.

Great Park Ice & FivePoint Arena is dedicated to ice hockey and skating.

Full-Circle Farm will present scores of crops and promote farm-to-table philosophies.

An arboretum will feature trees from across the globe, including oaks, palms, and other Mediterranean species.

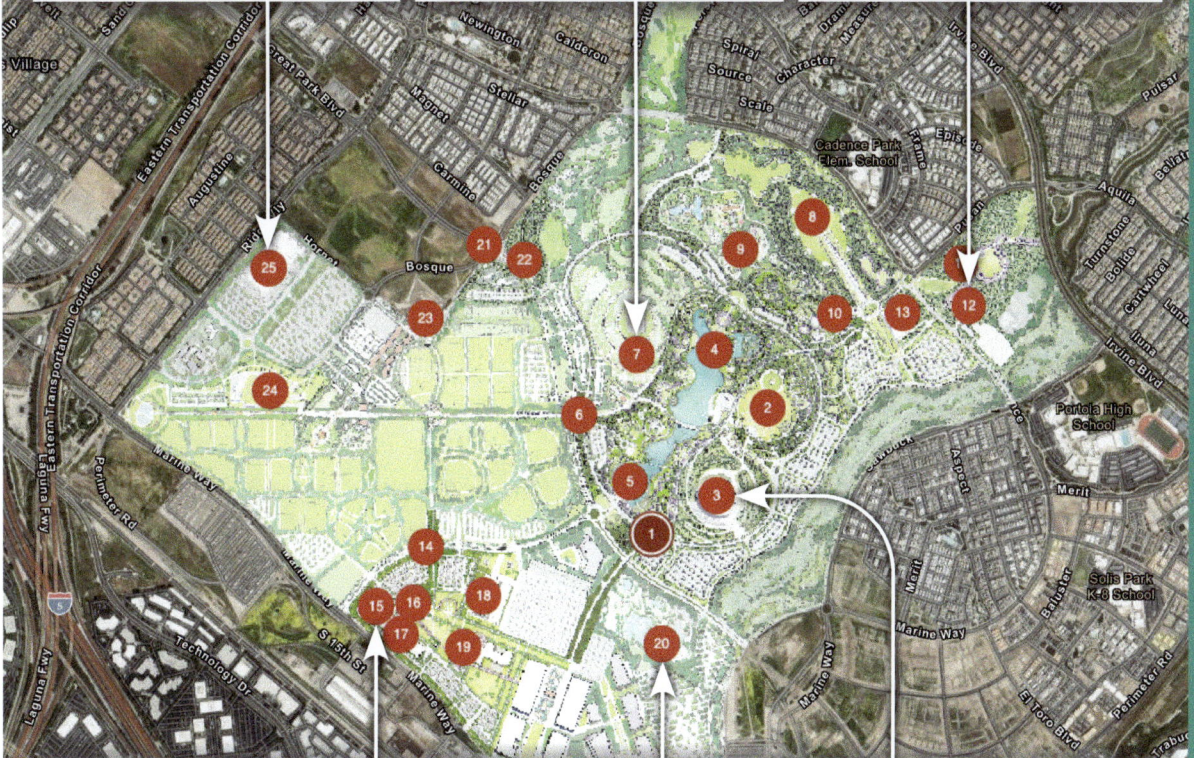

Pretend City, a children's museum, will open in 2026.

The Wild Rivers water park boasts 20 water slides and other attractions.

Plans call for an amphitheater that will seat 8,000-10,000 people.

# 6

## Maps in Dramatic Roles

NOAA works with national agencies, state agencies, and universities to model sea level trends and project future change around the world's coastlines. Red arrows represent rising sea levels; blue shows where sea level is falling.

# How maps perform within multimedia narratives

Over the years, I've created scores of multimedia narratives using the ArcGIS StoryMaps intuitive builder function. In a few of my stories, a map makes a single, modest appearance at the outset of the narrative, helping set the stage for a story that occurs within a single locality. In other cases, maps might appear multiple times in supporting roles. Example: a travelogue featuring several points of interest along a route, with each item meriting a map identifying its location. In still other stories, maps might star as the predominant element. A story about megacities, for instance, features the world's 10 largest metropolitan agglomerations, with each city represented by a sequence of maps showing its growth over time.

In describing how maps function within narratives, I've often found myself using theatrical terms and metaphors. In the previous paragraph, I mention *appearance*, *stage*, *supporting roles*, and *star*. The thespian terms may not precisely describe the various roles of maps, but the analogy is useful and revealing. The *stage* for a web-based multimedia story, for instance, is a web browser window or a smartphone screen. The *set* is analogous to the choices authors have made of type fonts, color palettes, and decorative elements. The *actors* are the photos, videos, audio—and, of course, maps—that appear within the story, or on stage.

In the theater world, some actors may be extras with bit parts. Others may serve as narrators. Still others may play supporting roles, or they may be the stars of the show. So, too, can maps, as *dramatis personae* in this realm of multimedia storytelling. Let's explore these various roles and describe some examples.

A globe locator.

Locator for *At Nature's Crossroads.*

## Maps as bit parts: Locators

Locators play a modest but important role in narrative dramas by providing geographic context. They're usually small and static. They can put a location in a global setting (*top left*) or use a familiar geography as an orienting device. The California map (*bottom left*) was created for a story called **At Nature's Crossroads** ⊝ about the Jack and Laura Dangermond Preserve, produced by The Nature Conservancy's California chapter in collaboration with my team and in celebration of newly protected coastal habitats near Santa Barbara.

Even the simplest of locators are typically the product of careful thinking and serial editorial judgments. Will readers recognize the shape of

California? Would it have been better to locate the preserve on a map of the United States? California's distinctive shape is probably recognizable to most readers, and adding the name of the state reduces the potential ambiguity. Had the preserve been in Colorado or Wyoming, with their generic rectangular shapes, the decision might have been different.

Locators should be as simple as possible. All items that might complicate their performance should be removed. In this case, I could have added San Francisco and Los Angeles to the map; I could have included shaded relief showing the Sierra Nevada and the Coastal Range. I could have added an outline to the state's shape, or perhaps even a scale bar. I could have enclosed the map within a box. But none of these items was necessary to the narrative or would have added value for readers.

A rule I try to live by in designing maps of all sorts is to **minimize the signal-to-noise ratio**. I ruthlessly eliminate elements that aren't essential to the story they're telling or the role they're playing. I especially like to represent areas—such as states—with flat tints instead of outlines, because line work almost always adds a visual complexity that's often unnecessary. There are plenty of circumstances where greater detail is required, but especially in the case of locator maps, less is more.

Multimedia storytelling offers another, more elaborate locator treatment, namely a scrolling experience that might be called a fly-in. The same Dangermond Preserve story features a good example. As readers scroll into the story, they encounter an immersive section that features a 3D web scene of the West Coast with a small green square that locates the preserve. As readers continue to scroll, the scene zooms in to closer views. In the final frame, the point of view tilts to an oblique angle, revealing the preserve's rugged topography. Fly-ins can be effective story overtures, providing geographic context and setting a scene.

Another bit part, less common than locators, is the comparison map. In the Dangermond Preserve story, the authors thought it would be helpful to give readers a sense of the preserve's size, since few have a sense of how big 24,000 acres really is. Although Manhattan is a continent away from the preserve, it's familiar to many and of roughly similar size.

Flying into the Dangermond Preserve.

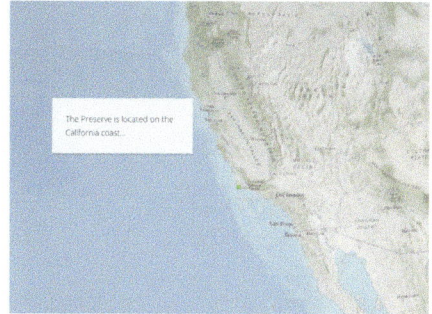

Think about how many judgment calls went into this simple graphic. Not just the choice of place to compare the preserve to, but the positioning of the elements, the colors, the transparency, the type style, and letter spacing. Note how "Manhattan" is partly outside the preserve boundary to make clear which feature is being labeled. This sort of attention to detail helps make a map perform its role most effectively.

## Maps as narrators

Sometimes, maps within stories serve to locate things, but they play a more active part in the production, something akin to a narrator.

A standard and often-used feature of ArcGIS StoryMaps is the map tour immersive, which features a sequence of geolocated photos or videos. The maps place items within a landscape, locating each item relative to the others. Readers can navigate sequentially through the tour by scrolling or clicking on pointers, but they can also interrupt the narrative by panning and zooming on the map and clicking icons representing other items in the tour. This story about endangered World Heritage Sites uses the map tour immersive.

The narrator metaphor is especially apt here because the maps stand rather modestly off to the side of the images and text, not unlike the way a sign interpreter is situated at the edge of the stage during a theater production. This story uses the media-focused map tour layout.

In 2023 our team collaborated with the National Oceanic and Atmospheric Administration (NOAA) on **Coastal Flooding** ⊂⊃, a story exploring

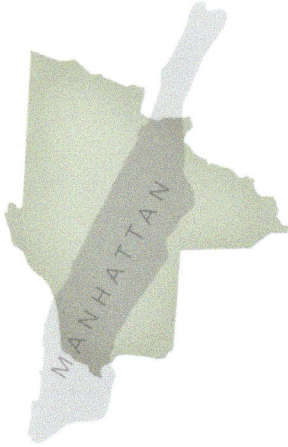

Comparing the Dangermond Preserve to Manhattan.

Maps as narrators: World Heritage Sites in danger.

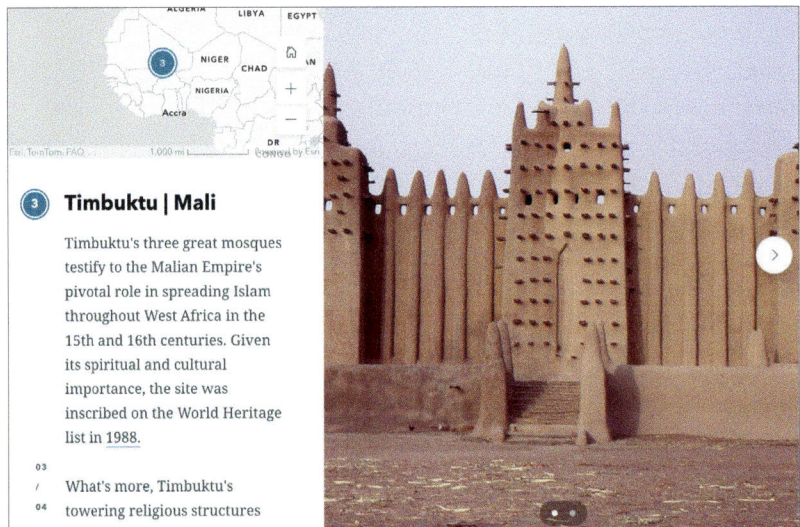

### Timbuktu | Mali

Timbuktu's three great mosques testify to the Malian Empire's pivotal role in spreading Islam throughout West Africa in the 15th and 16th centuries. Given its spiritual and cultural importance, the site was inscribed on the World Heritage list in 1988.

What's more, Timbuktu's towering religious structures

**New Orleans, Louisiana, USA**

The city of New Orleans is located on the Mississippi river near the Gulf of Mexico.

In 2005, Hurricane Katrina hit the Louisiana coast, causing the water level to rise nearly 15 feet in some areas. This storm surge destroyed houses and infrastructure and caused billions of dollars of damage.

*Photo: NOAA*

**Kiribati**

It is predicted that millions of people living on low-lying islands in the Caribbean Sea and the Indian and Pacific Oceans will face displacement due to sea level rise and coastal flooding.

The entire Pacific island nation of Kiribati may relocate in the next decade due to sea level rise. This small Pacific island nation is trying to cope with contamination of freshwater reserves and food crops from coastal flooding.

**2060 Sea Level Rise**

By the year 2060, we see that the number of structures at least partially below the high tide line has significantly increased and that most of downtown Norfolk is impacted by some form of flooding.

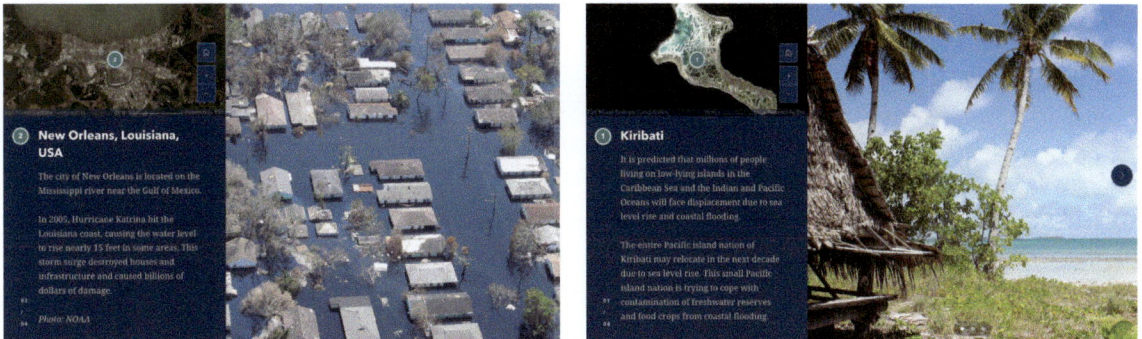

*Above*: A tour of locations where concerns about sea level rise are acute includes stops at New Orleans, Louisiana, and Kiribati in the Pacific Ocean.

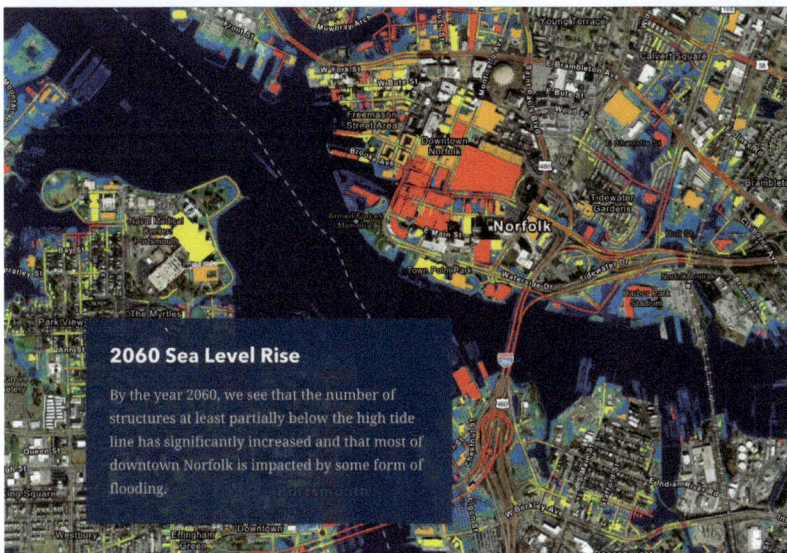

*Left*: A map of Norfolk, Virginia, highlights locations predicted to be impacted by coastal flooding.

the impacts of climate-change-induced sea level rise on coastal communities. The maps within the story perform several roles, including narration. A tour takes us to four corners of the world where coastal flooding is a particularly acute concern—including New Orleans and Kiribati, an island nation in the Pacific. A small "narrator" map modestly places each location in a geographic context, leaving most screen real estate to images and text.

By the way, "Kiribati" is pronounced KEE-ruh-bas. Why do I know? During my Nat Geo days I was invited to appear on the Today show to describe where on the globe the sun would first rise on January 1, 2000. Having done my homework, and as I told Al Roker, the International Date Line jogs eastward to encompass the whole country of Kiribati, providing it with an early sunrise and temporary notoriety.

Elsewhere in the coastal flooding story, maps play a more active role, predicting future flooding in Norfolk, Virginia. The story contains a larger lesson. The maps throughout the story are interactive; readers are welcome to dive into the maps by panning and zooming. But we don't insist that readers explore the map. We don't even explicitly invite them to do so. (The maps' pan-and-zoom controls are tucked into the lower right corner, giving readers a hint that there's optional exploration to be done.) Instead, we designed the story so that readers can learn simply by scrolling. Research has consistently found that most readers don't bother to interact with maps by manually panning, zooming, and clicking around. Doing so can quickly become tedious, and it distracts readers from the story's narrative flow.

But interactive maps can be fun to explore, providing additional information for the subset of readers who are eager to learn more. In addition, GIS professionals make up an important part of our storytelling community. GIS users have experience interacting with maps, and most of them love maps. So I like to include interactive maps as a kind of bonus feature, usually without forcing readers to use them.

Narrator maps can feature a series of points along a route, of course. One of my favorite use cases for this approach is a product of a long-term collaboration with Paul Salopek, a Pulitzer Prize–winning journalist and National Geographic Explorer. Paul is walking across the world, tracing the human diaspora from its African origins to its farthest-flung extent in South America's Patagonia region. Along the way, Paul is pausing at 100-mile intervals to record milestones, for each of which he adds a brief description, interviews a passer-by, takes photos of the ground at his feet and the sky above, and records a video of the scene, no matter how banal.

Our team's developer, Lee Bock, created a custom story, **Out of Eden Walk Milestones** ⊂⊃, with an interactive map and side panel that features thumbnails for each milestone. Clicking a thumbnail or map icon expands the side panel and presents the milestone elements. Readers can navigate the story by clicking map locations or previous and next pointers in the side panel.

Like the story about expanding airports, readers can click sequentially through the locations using the panel controls or wander around the interactive narrator map at random. Providing these alternative means of navigating serves casual readers and more dedicated and curious visitors alike.

Individually, Paul's milestones are mildly interesting. Collectively, they portray the rhythm and variety of Paul's trek and the mix of ordinary and extraordinary that every location possesses. That mix becomes more apparent as one browses through the milestones in sequence. It's that

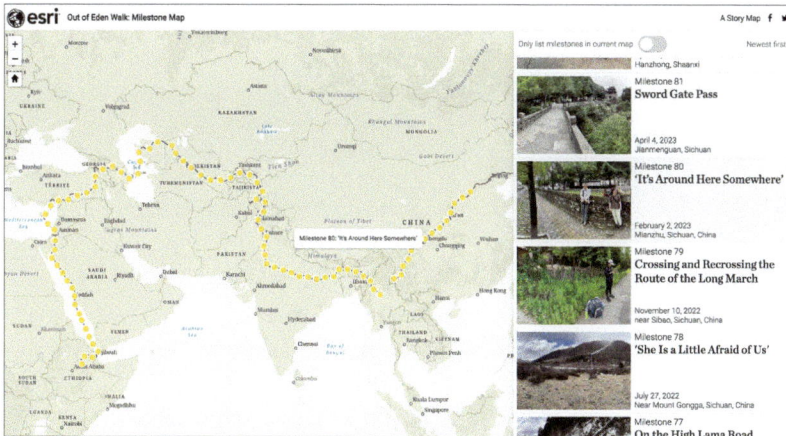

*Out of Eden Walk Milestones.*

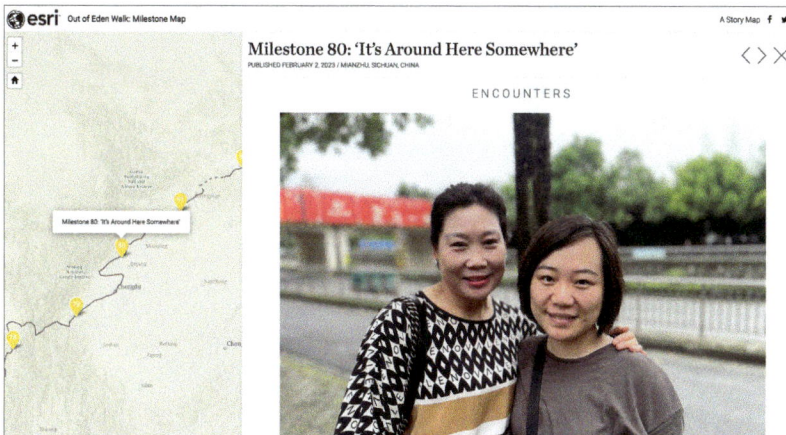

juxtaposition of the quotidian and the revelatory that Paul explores with passion and sensitivity in his slow journalism—and that we experience but all too rarely relish—in our daily lives.

Another variation on the narrator map theme is the index map, where an interactive map near the beginning of a multimedia narrative links to the components of a virtual tour. The map can be thought of as the cartographic version of a table of contents. An example is a story created by my team on **Abandoned Islands** ⊖ around the world. The main body of the story is a sequential tour of a dozen islands, all of which have colorful histories of settlement and desertion. One way to navigate the story is to simply scroll through it and explore the islands sequentially. Another is to point your cursor to icons on the index map representing each island, and then click pop-ups to jump to their descriptions within the story. At the end of each island's entry is a Return to Map link that sends readers back to the index map. It's an effective technique in this instance, since each island

# ABANDONED ISLANDS

For a variety of reasons, these islands have vacancies.

**By Esri's StoryMaps team**

Planning a relaxing beach getaway? You'll probably want to steer clear of these 12 uninhabited islands. Some of these islands were depopulated due to insurmountable environmental forces (an infestation of venomous vipers, for example), some due to economic forces (changing shipping patterns), some due to

A tour of abandoned islands around the world.

Click the points above to jump to an abandoned island.

## SUAKIN ISLAND, SUDAN

For nearly three millennia, the small island of Suakin, located on Sudan's Red Sea coastline, served as a strategic port for the great empires of region. First charted in the 11th century BCE by envoys from the court of Pharaoh Ramses III, Suakin prospered into a wealthy commercial city over the next several hundred years, and at its height was known throughout the Old World as a symbol of medieval wealth and luxury. All of Suakin's buildings were constructed from glimmering coral, and embellished with detailed stone and wooden carvings depicting its historical glory.

The weathered ruins of the ornate structures are all that remains of Suakin Island.

has its own separate history. It would not be appropriate for narratives in which locations need to be seen and read in sequence.

## Maps in supporting roles

Because Esri is a mapping company, and because we love maps, we assumed early in our team's tenure that interactive maps would be ubiquitous within ArcGIS StoryMaps, and that other multimedia content would be confined to secondary status. But we came to realize that there are countless stories for which maps can add a valuable, even essential, dimension without dominating the narrative.

The Amazon Conservation Team nonprofit works with Indigenous groups in the upper Amazon basin to protect the region's rich biological and cultural diversity. The group has produced many outstanding multimedia stories over the years; my favorite is **Living Territories** ⟜⟶, which makes full and creative use of an abundant mix of maps, images, audio, and artwork.

A sampling of screenshots (*next page*) from the story gives a sense of its visual beauty and variety and the degree to which maps are intimately intermixed with text, photos, and illustrations. The story opens with a series of full-screen images that vividly sets the scene and introduces Indigenous people into the narrative. Thematic maps, most of them relatively simple and easily interpreted, provide context by revealing the extents of Indigenous territories and delineating the threats they face from deforestation and illegal mining. Some of the maps directly incorporate artwork; the authors used the capability of ArcGIS Online, Esri's vast Web GIS resource, which enables the addition of images as media layers within web maps. Portraits and audio clips humanize the narrative and underscore the Amazon Conservation Team's commitment to working with local groups to protect cultural and biological diversity.

Another fine example of maps in supporting roles is **Justice Deferred** ⟜⟶, an account of the system of internment camps within which thousands of Japanese Americans were confined during World War II. The story was produced by the ArcGIS StoryMaps editorial team and features elegant maps by Cooper Thomas, a cartographer on the team.

Near the top of the story, well before readers encounter any maps, is an array of portrait photos of Japanese Americans. This human touch reminds us that the statistics charted and mapped later in the story profoundly affected tens of thousands of American citizens and immigrants who were uprooted from their homes, schools, and jobs because of their ancestry. It's

A sampling of screens from Amazon Conservation Team's story *Living Territories*.

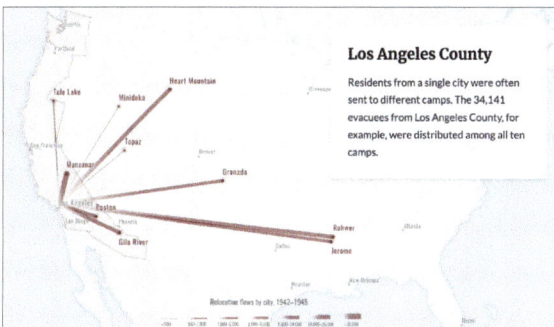

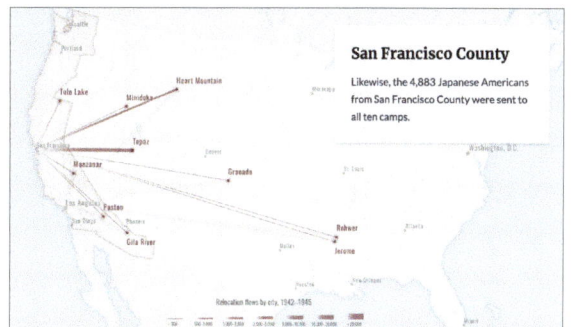

an important lesson: Maps can be dispassionate; pairing information-dense cartography with emotion-dense imagery adds depth and nuance to multimedia narratives and stirs us, as readers, to empathize with the humans whose lives were disrupted by historical events.

While doing research for the story, Cooper digitized data on camp populations that he had discovered in old records. He brought the data to life in a series of maps within the story. An unexpected result of his work is that the data has since been cited, reused, or discussed dozens of times by other publications.

The *Justice Deferred* story is told in six clearly organized sections: the prelude, the order, the relocation, the camps, the return, the legacy. Its

*Justice Deferred*, featuring World War II-era portraits of Japanese Americans and maps depicting the relocation of Japanese American populations to internment camps.

intermingling of period photos and new and archival maps paints an informative and affecting picture of a shameful episode in American history.

## Maps as stars

Then there are stories in which a map (or a series of maps) is the primary actor on the stage, whereas other multimedia content, if it's present, plays a far less dominant role in the production. Starring maps can play a variety of roles.

They can appear as a series of thematic maps of the same geography. Among my favorites in this category is **Mapping the Thanksgiving Harvest** ⊖. It visualizes data from the US Department of Agriculture documenting, for every US county, the number of acres devoted to growing specific crops and the number of farms growing them. The story charts eight foodstuffs that are likely to appear on Thanksgiving dinner tables: cranberries, turkeys, sweet potatoes, potatoes, green beans, brussels sprouts, pumpkins, and pecans. It employs a single, nonessential photograph, of cranberries, on its title panel. Each map is accompanied by a brief paragraph of interpretive text.

*Mapping the Thanksgiving Harvest*: cranberries (*red*) and pecans (*brown*).

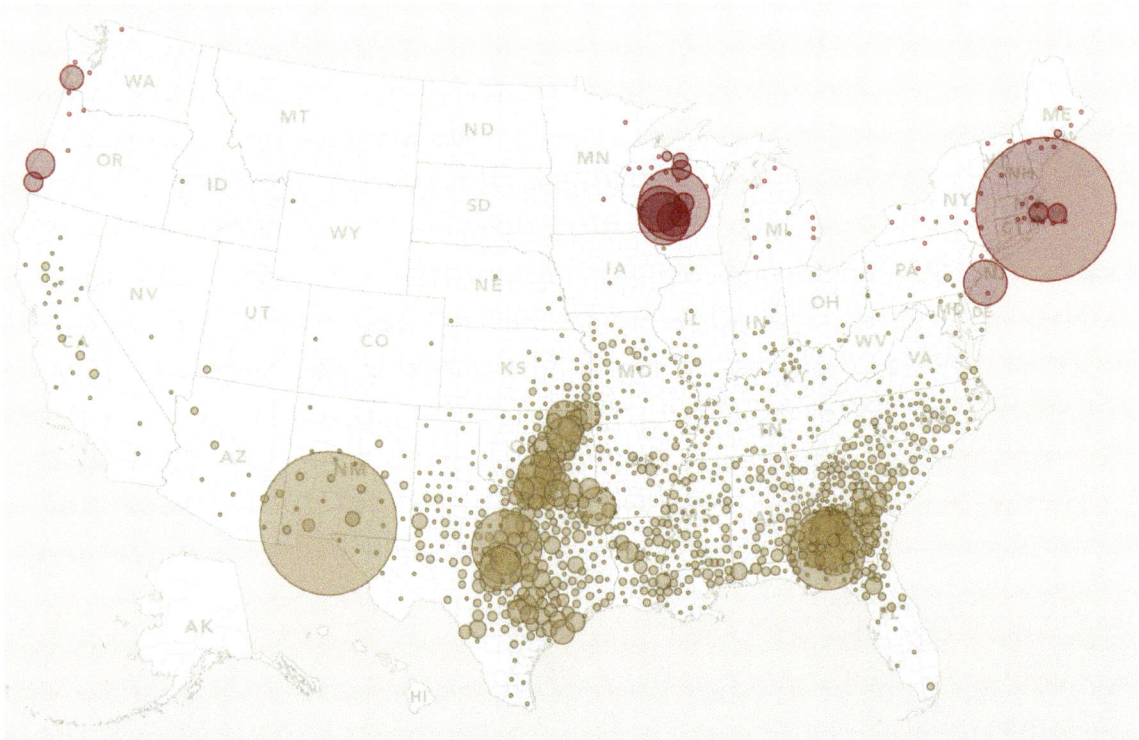

The maps convey the obvious message that the foods you consume at your Thanksgiving feast come from very different parts of the country. I've combined two of the maps here, because there's virtually no overlap between counties where cranberries are grown—mostly in moist places along the country's northern tier—and the more southern distribution of pecan farms. The implicit message in the contrasting maps is the infra-structure required to distribute the harvests from these disparate sources to your local supermarket and thousands of others—an everyday miracle that we take almost entirely for granted.

Multimedia stories can also feature a set of similar maps of different places, enabling comparisons among locations. The **Age of Megacities** ⊖⊃ story is among the most viewed of the stories my team has produced. It has an all-star cast of maps as actors: Near the beginning of the story is an attention-grabbing 3D map animation showing a spinning globe bristling with spikes representing urban populations. But the heart of the story is a succession of maps (*next page*) for 10 of the world's largest urban concen-trations, showing the increasing extent of their built-up areas over time. Readers click buttons for each period. The buttons activate what we call media actions. Buttons can be configured to show or hide individual layers in a web map.

Counting these map actions, the story presents some 50 maps. That's a lot of maps. What makes it digestible is that it's consistently organized, with the same pattern repeated for each city.

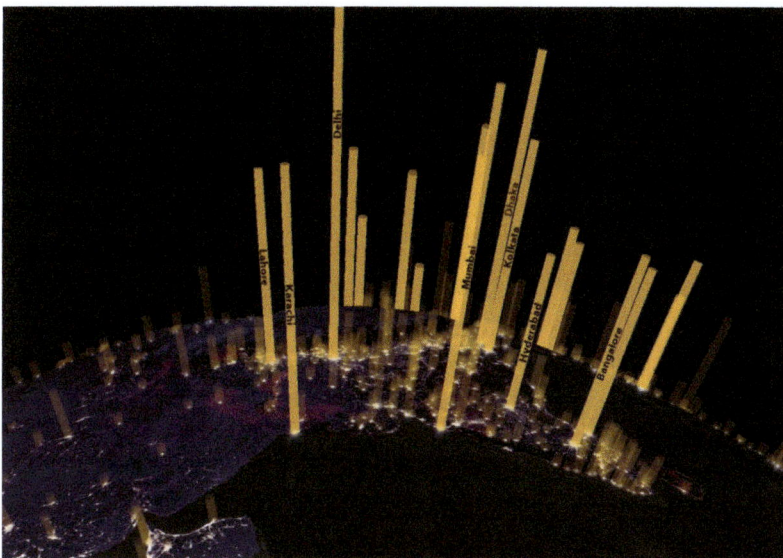

Image from a 3D anima-tion of world cities from *Age of Megacities.*

Collectively, the city maps illustrate the dramatic shift of the world's population from rural to urban. Closer examination reveals other patterns: For instance, the slower growth of the built-up areas of London and Paris reflect Europe's slower population growth versus the explosive growth of Lagos, Nigeria, and Mexico City.

Some of the most effective stories establish a *rhythm*—a repeated pattern or sequence. Readers quickly become accustomed to the rhythm and gain an intuitive understanding of the structure of the story. Repeat the rhythm too many times, though, and you risk boring your readers. That's why we limited our urban tour to 10 cities.

Profiling megacities. *Left*: São Paulo's urban extent in 1905, 1949, and 2020. *Right*: Tokyo in 1929, 1954, and 2020.

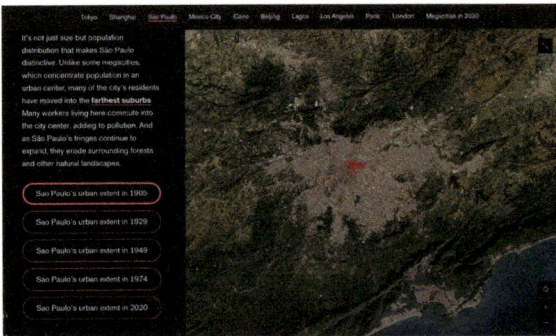

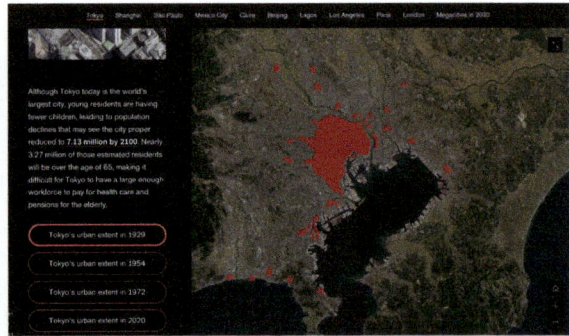

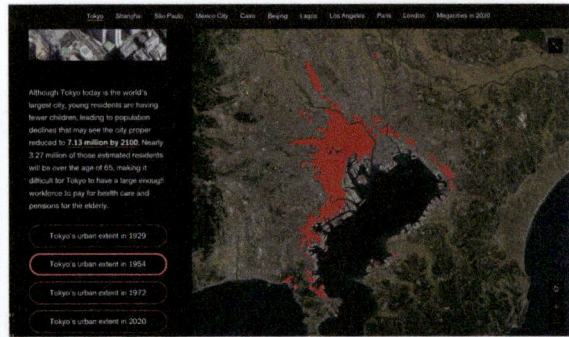

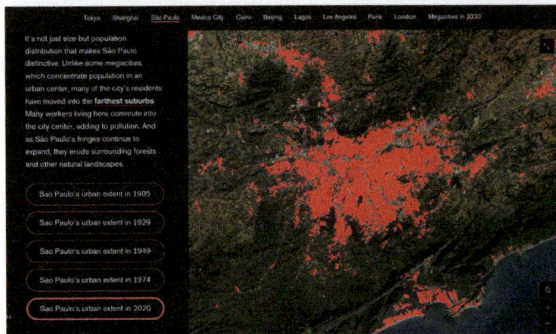

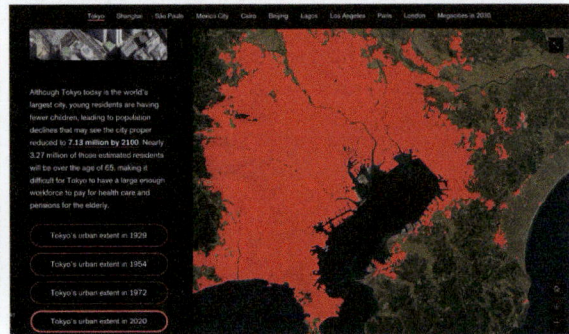

## Acting techniques

It's not just the type of role a map plays in a narrative, of course. It's the style of acting. Maps, like actors, have different performing styles. Some resort to histrionics; others are understated. Some demand attention and require interaction. Some play it cool, letting the narrative predominate rather than hogging the spotlight.

Let's say we're telling a story about the geographic dimension of income inequality in US cities, as we did in **Wealth Divides** ⊖. We could either let you, the reader, explore on your own, or we could take you on a guided tour. The acting metaphor starts to fail us here, but a fully interactive map might be a bit like an improvisation, and a guided tour could be akin to a Shakespearean soliloquy.

We could present you with a map of the United States and ask you to pan and zoom to a city of your choice, and then explore neighborhoods with high and low incomes. Or we could take you directly to, say, New York City, and ask you to explore within it, clicking census tracts for data.

But we thought it might be more effective if we introduced the issue of income divides by highlighting a tract on New York's Upper East Side and informing readers, as they scroll, that median household income in this area is

*Wealth Divides*: illustrating a national issue with local examples. Note the use of a similar color palette for title art (*below*) and maps.

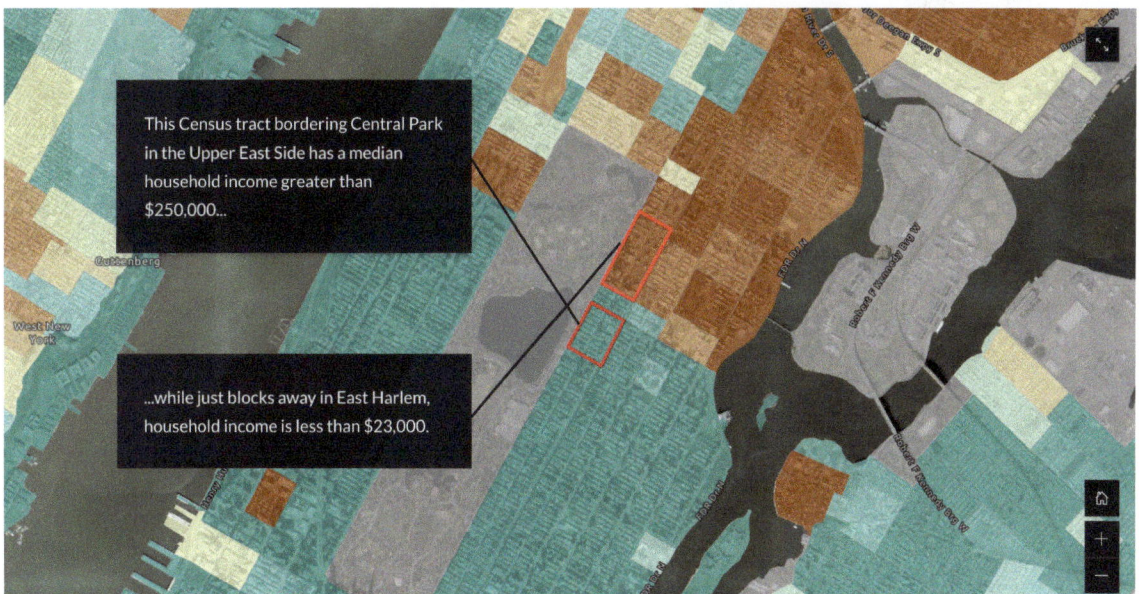

more than $200,000, while less than a mile away, in a South Harlem tract, income is below $16,000.

We took readers on similar guided tours of income divides in Atlanta, Boston, Washington, DC, and San Francisco.

As directors of a play, or producers of a multimedia narrative, we can choose to let the actor—the map—steal the spotlight. Or we might urge the actor toward a subtler performance that better supports the narrative. Residing within a US census map of household income are myriad stories. By guiding the reader to a small number of tracts representing income extremes in close physical proximity, we're revealing a particular story that we think is compelling. And we're highlighting a specific, local example that represents a much larger phenomenon.

I'm willing to bet that most casual readers won't mind being led rather passively through a story if the story itself is engaging. One can strive to provide the best of both worlds by maintaining an engaging story line while giving readers the option of exploring on their own—which is exactly what we did in the *Wealth Divides* story.

Interactive maps can be fun to explore, and it's empowering to create and publish interactive maps. But in many—perhaps even most—cases, interactivity is unnecessary at best, a distraction at worst. As director of a play, and as producer of a multimedia narrative, I want all the players on my stage to support the story line. Individual quirks and mannerisms might be acceptable, but only if they help engage the audience and rein-force the message I'm seeking to convey.

## Maps as bad actors

Although hundreds of thousands of ArcGIS StoryMaps stories have been published, we've encountered very few abuses of the medium. Among the few violations of norms or laws we've encountered, the majority have to do with authors' use of images or other media without securing the proper permissions. The ArcGIS StoryMaps editorial team is conscientious about securing written permission for all elements of the narratives we publish—and we hope others follow our example.

The other category of potential abuse is cartography. Maps can become bad actors if their authors make choices, intentionally or innocently, that mislead. People tend to believe maps; they have a look of authority that may not be entirely justified. For more than three decades, Mark Monmonier's book *How to Lie with Maps* has provided excellent guidance on mapping best practices.

A 2021 panel discussion with Jack Dangermond, journalist and former *Atlantic* contributing writer James Fallows, and former National Geographic Society president and CEO Gary Knell raised the question of ethics in mapping. It led to an effort by Esri research cartographer Aileen Buckley, director of product engineering Clint Brown, and me to draft a "Mapmaker's Mantra" to provide guidance, in easily digestible form, to makers of maps. Its elements:

- **Be honest and accurate.** The highest objective and primary obligation of ethical mapmakers is to communicate information in the most accurate and understandable way. They strive for veracity and verifiability in all aspects of their mapmaking.

- **Be transparent and accountable.** Ethical mapmakers take responsibility for their work and are open and transparent about their sources and decisions. They accept that neither speed nor format forgives accountability.

- **Minimize harm and seek to provide value.** Ethical mapmakers treat sources, subjects, colleagues, and members of the public with respect; they promote equity, inclusion, and empathy. They strive to make maps of value to increase understanding and provide insights.

- **Be humble and courageous.** Ethical mapmakers humbly admit when they get it wrong and gently point out when others get it wrong. They have the courage to admit when they do not know something and call on others when their own skills and knowledge are insufficient.

It's the responsibility of all of us—especially in this age of misinformation—to be responsible and truthful storytellers, not only in the maps we make but in all the elements of the narratives we create.

# Dante's *Inferno*

## According to A. Manetti & G. Galilei

**Author:** Josef Münzberger

**Medium:** ArcGIS StoryMaps

**Story behind the story:** Josef Münzberger, formerly a graduate student at the Czech Technical University of Prague, used GIS tools to map the circles of hell as described by Dante Alighieri in *Divine Comedy*, a narrative poem he composed in the 14th century.

**Why it's special:** About a century after *Inferno* was published, Antonio Manetti and Galileo Galilei used place-names mentioned in Dante's text and an estimation of the size of the earth to plot the location of hell and its various circles. As Münzberger explains in a paper published by the International Cartographic Association, hell "resembles an enormous conical abyss with its vertex in the center of the Earth divided into certain levels in which various sins are punished."

Dante's circles can be visualized as three-dimensional, descending in a series of nested cylinders and coming to a point at the earth's center, as shown in the 3D model.

**The author:** Münzberger summarized his master's thesis in a lively multimedia narrative that describes what may be the only GIS-based map of hell.

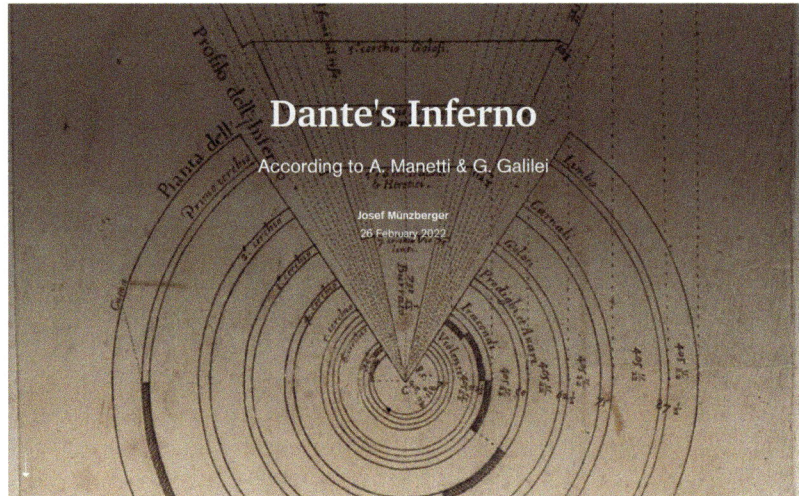

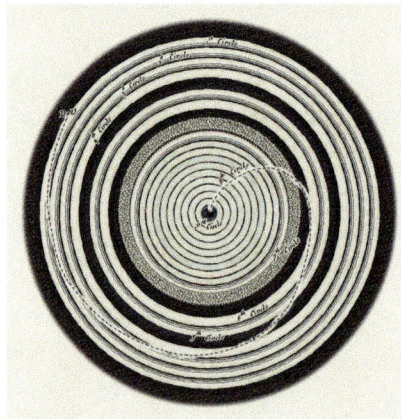

*Top*: Cover art and title.

*Above*: Each circle is related to a particular sin. The worst sin, according to Dante? Treason.

*Right*: The circles of hell visualized in three dimensions.

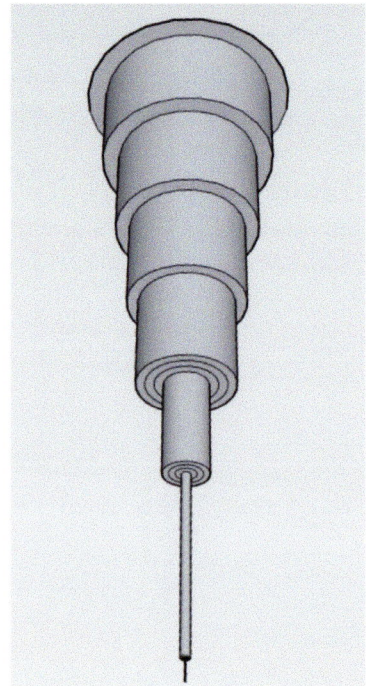

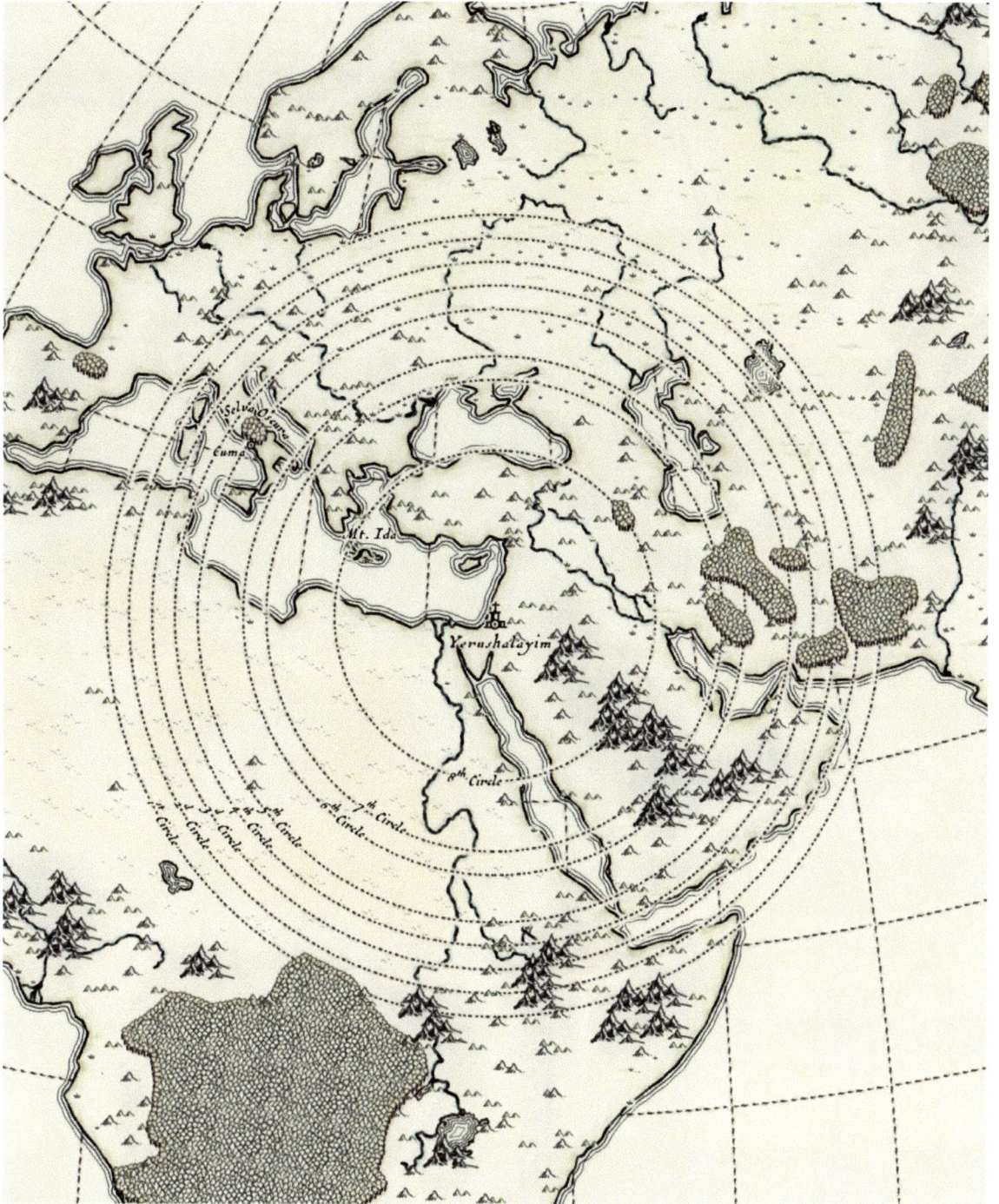

Hell's outer regions mapped as concentric circles centered on Jerusalem.

# Okavango Explore

Using EarthViews maps to understand one of the planet's most important waterways

**Author:** EarthViews and Blue Water GIS

**Medium:** ArcGIS StoryMaps

**Story behind the story:** "Transport yourself to this place of mystery and richness," suggest the authors of this visually stunning narrative. The story paints a picture of one of the world's most distinctive ecosystems, an inland river delta fueled by a natural "water tower" in the sparsely populated highlands of Angola. Seasonal waters flow from the highlands into Botswana and spread out into extensive wetlands that harbor abundant wildlife.

**Why it's special:** The story portrays a series of expeditions organized by the National Geographic Okavango Wilderness Project; portions of the story include EarthViews 360° images that field scientists used to document natural environments.

The Okavango region faces multiple threats, including resource extraction, water diversion, wildfires, poaching, and deforestation. The story helps raise awareness about a little-known ecosystem in urgent need of protection.

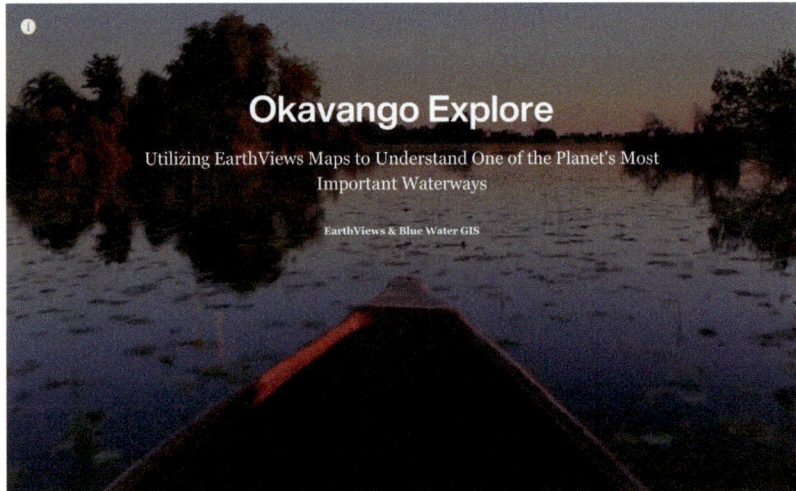

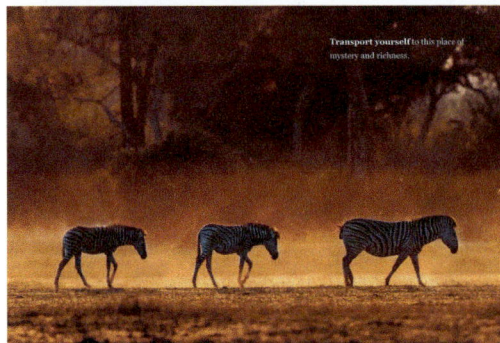

## Okavango Explore

Utilizing EarthViews Maps to Understand One of the Planet's Most Important Waterways

EarthViews & Blue Water GIS

Transport yourself to this place of mystery and richness.

The story opens with a video (*top*) that places readers in the midst of the natural environment. Photos portray the region's biodiversity (*left*).

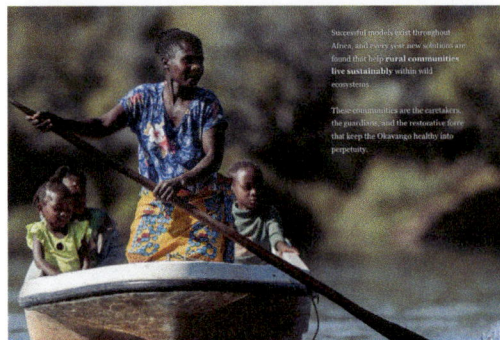

Successful models exist throughout Africa, and every year new solutions are found that help rural communities live sustainably within wild ecosystems.

These communities are the caretakers, the guardians, and the restorative force that keep the Okavango healthy into perpetuity.

*Left*: The story acknowledges the local Indigenous people and stresses the importance of their participation in protecting the Okavango landscape.

Most of the water that sustains the richness of the Okavango Basin has its origin in the semi-humid Highlands of Angola, and it completes its downward journey in the lowlands of Botswana.

Rising in southeast Angola, the **Highlands** capture precious rains and serve as the Okavango Delta's **Water Tower**, an area of lakes and aquifers that **supplies nearly all of the water** to the Delta downstream.

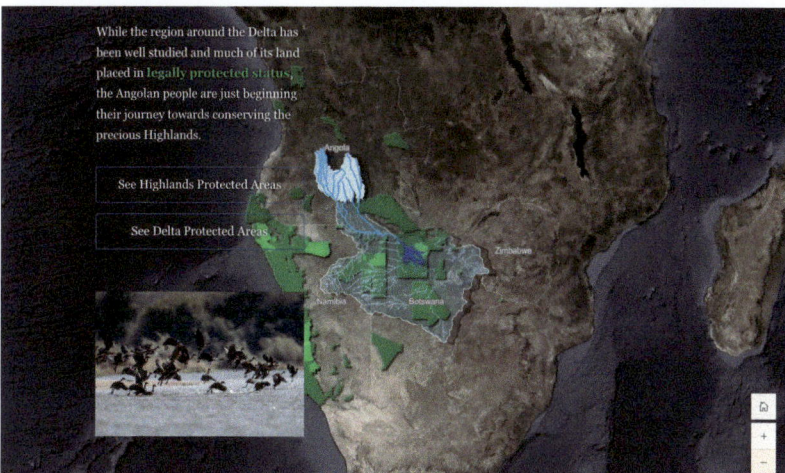

While the region around the Delta has been well studied and much of its land placed in legally protected status, the Angolan people are just beginning their journey towards conserving the precious Highlands.

See Highlands Protected Areas

See Delta Protected Areas

A sequence of choreographed maps locate the Okavango River Delta and introduce readers to its hydrology (*top*), seasonal cycles (*center*), and protected areas (*bottom*).

Urumqi

*Gobi Desert*

Baotou

Tai

Xining

Lanzhou

Xi'an

CHINA

Xiang

Yichang

Chengdu

Chongqing

*Plateau of Tibet*

*Salween*

*alaya*

Thimphu

Guiyang

Guwahati

Lijiang

Patna

Kunming

Liuzhou

Nanning

*Mandalay*

Hanoi

Zhanjiang

*Gulf of*

# 7

# Nine Steps to Great Storytelling

Qiqihar

Daging

Harbin

Manchurian Plain

Changchun

Hegang

Shuangyashan

Jixi

Mudanjiang

Vladivostok

Anshan

gjiakou

Beijing Tangshan

aoding

Dalian

NORTH
KOREA
Pyongyang

Yantai

Jinan Weifang

Qingdao

Taizhou

Hefei

Shanghai

Hangzhou

Wenzhou

Fuzhou

Quanzhou Taipei

TAIWAN
Kaohsiung

# Why nine steps?

Throughout my 27 years at the National Geographic Society, I enjoyed the company of some of the world's best storytellers—writers, photographers, illustrators, videographers, and, yes, cartographers. The society attracted talented storytellers like honeysuckle attracts bees. Storytelling, with images and prose, was our stock in trade, although we only rarely referred to ourselves as storytellers. The term had yet to enter the vernacular to the extent it has saturated the zeitgeist since the turn of the millennium.

A particular trait that many of us at Nat Geo did *not* possess was technical aptitude. We could weave a tale, but for the most part, we couldn't write code or program a computer, nor were we particularly interested in learning how to do so.

Fast-forward to my move to Esri in late 2010 and our early efforts to provide storytelling tools to Esri's core audience of GIS professionals. As we developed our story-building tools, it slowly dawned on me that we were serving a community that was quite comfortable with technology but relatively clueless about story*telling*. Our users were quick to master the nuts and bolts of our early story builders. But, with notable exceptions, few of them knew the basics of how to capture the attention of readers, keep them engaged in a well-structured narrative, and, ideally, inspire them to action of some sort.

Early on, our primary communication with our audience focused on which buttons to click, which formats to use, how to incorporate maps, and so on. Our users were eager for this information, but I was slow to realize that they were at least as eager for some basic lessons in spinning a memorable yarn.

Over the years, my team has produced scores—perhaps hundreds—of multimedia stories, ranging from modest instructional pieces to ambitious

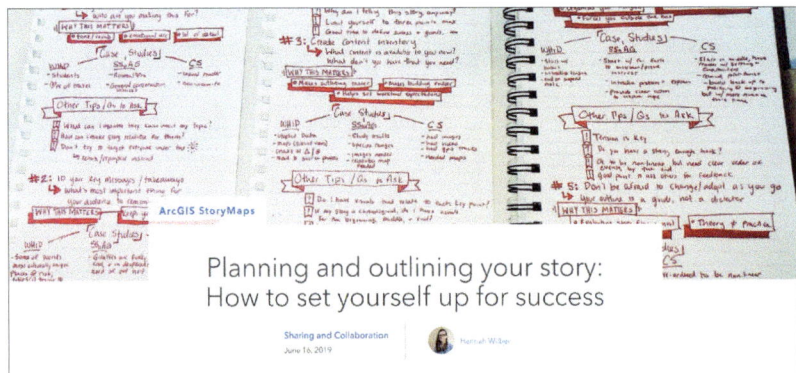

Title panel of Hannah Wilber's popular blog post on planning and building stories.

narratives, many created in collaboration with partner organizations. We've also perused thousands of stories produced by our global community of storytellers. Thus, we've come to recognize the elements that make a story effective.

A perennial favorite among our users is a 2019 blog post—"Planning and Outlining Your Story: How to Set Yourself Up for Success" by former team member Hannah Wilber. Meanwhile, I created an instructional story called **Nine Steps to Great Storytelling** ⊂⊃, which has also become a go-to resource for people seeking basic storytelling advice. It would no doubt have been possible to come up with 90 storytelling tips, but we wanted to limit our advice to a few broadly useful items and to complement the "Planning and Outlining" blog with advice on literally building your stories. The remainder of this chapter will take you through these nine steps and bolster them with examples of each item in action.

## 1. Start with a bang

How you begin your story makes all the difference. Our counsel: Don't ease into your story. You should make every effort to grab your readers by the lapels with a strong image—and a strong title. Better yet, make the title and image work together to set the tone for the story and pique your readers' interest. We like titles that aren't simply labels. They might include a verb. They might be alliterative. They might offer a play on words. Ideally, they echo—and help establish—the style and mood of the story, as well as letting your readers know the gist of what's coming.

Story covers: striving for strong titles and evocative images.

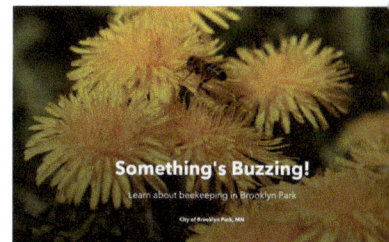

On foot in the path of the Silk Road
A walk through the birthplace of globalization
Paul Salopek | Out of Eden Walk

**Hot Numbers**
It's a fact: Human impacts are warming our climate and changing our planet

**Something's Buzzing!**
Learn about beekeeping in Brooklyn Park

National Geographic Explorer Paul Salopek chose the evocative pairing of image and title shown on the previous page to open his account of trekking along portions of an ancient trade route. The image perfectly complements the poetic and evocative title, **On Foot in the Path of the Silk Road** ⤵. It's hard to resist scrolling down to learn more about Paul's walk through the arid and mountainous landscapes of Central Asia.

**Hot Numbers** ⤵ is the title I came up with for a story summarizing aspects of the climate crisis by citing and explaining salient statistics. Is the title too cute, or too clever, for its own good? I worried about it at the time, but I've since decided that it works. Better something catchy than a title that scares people away from a difficult and emotional topic.

**Something's Buzzing!** ⤵—the title chosen by the City of Brooklyn Park, Minnesota, for a story on beekeeping—is unabashedly cute. But the title is appropriate to the light tone of the story, and it perfectly complements the looping video of a busy bee that forms the background of the title page.

Few multimedia narrative title pages evoke the surprise and (we hope) delight that our team's story on the spotted lantern fly did, an invasive insect that's spreading rapidly from east to west across the United States. Warren Davison came up with the brilliant notion of having a lantern fly crawl across the screen of **What's That Bug?** ⤵. He created an animation, made convincing by the research he conducted into the mechanics of six-legged locomotion.

You may have noticed that we're not showing examples of maps on story covers. There's a reason for this. As much as we love maps, we feel that images or short videos evoke a more immediate, visceral response. Maps are wonderful within stories, but for the most part they don't have the immediate impact that images do. Most maps also beg for an explanation of some sort; diving into that level of detail at the beginning of a story dilutes the impact that's so important to attracting readers.

ArcGIS StoryMaps offers several cover formats, including a split-screen effect with title and image side by side. I usually prefer the greater visual punch of a full-screen image, but other treatments can be effective. The next story, about the devastating effects of a winter rainstorm in the Russian Arctic, juxtaposes a photograph of a reindeer herd with a provocative title, **When Rains Fell in Winter** ⤵, to create an enticing invitation to learn more.

The title page also works well because its subtitle provides concise information about the story and effectively complements the poetic nature of the title: *A decade ago, heavy winter rains washed over the Yamal Peninsula*

## What's that bug?

Invasives are a big problem; a new insect could be an even bigger one.

By Esri's StoryMaps Team | December 12, 2023

A lanternfly crawls across the cover of a story on invasive species.

## When Rains Fell in Winter

A decade ago, heavy winter rains washed over the Yamal Peninsula in Northwest Russia, killing 60,000 reindeer and ruining livelihoods.

Philip Burgess & Irina Wang
March 9, 2023

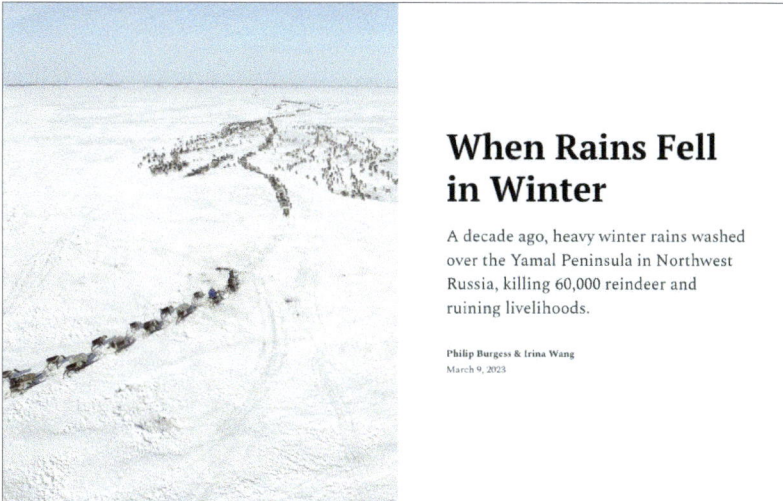

A wintry Russian landscape and an evocative title tempt readers to scroll.

*in Northwest Russia, killing 60,000 reindeer and ruining livelihoods*. The title evokes and provokes; the subtitle informs.

A caveat: Great design often involves the judicious breaking of rules. So take my authoritative statements with a grain of salt. Is there ever a situation where a map might be the best possible treatment for a story cover? Almost certainly. But note the qualifier: *judicious* breaking of rules. You should have a thorough familiarity with the rules of good design before violating them!

## 2. Add a hero

People love people. Nearly all of us love to look at people, and we love to read about them. It's much easier for us to identify with people rather than abstract issues. Adding a human face to your story helps your readers build a bridge between an individual and a larger con-text or issue. By no means do all stories lend themselves to adding a main character, but many do.

Your hero may be a volunteer, a scientist, or a historical figure. Or it may be a more or less ordinary citizen who is affected by a larger phenomenon. That's the case with Dorothy and Nathan, friends of team member Will Hackney. Will was working on **Charging Across the Country** ↪, a story about electric vehicle charging stations, a network of installations that is growing rapidly but that continues to have gaps and limitations that can make motorists' knuckles whiten. Will describes a weekend outing by Dorothy and Nathan to a Delaware beach resort, and the couple's frustration as a shortage of available stations forced an inconvenient change of plans.

Their personal experience drives home the fact that a nationwide map of

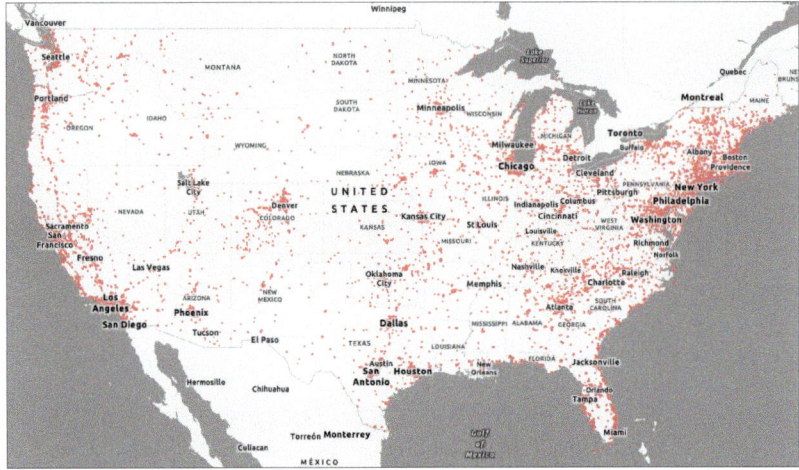

U.S. charging stations.

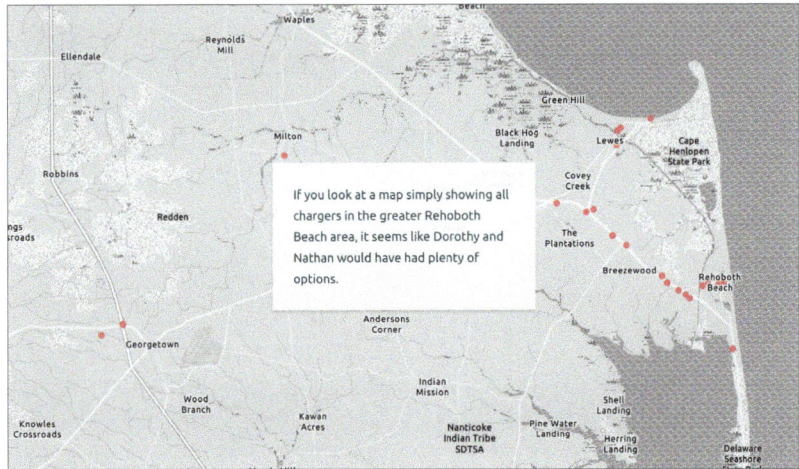

If you look at a map simply showing all chargers in the greater Rehoboth Beach area, it seems like Dorothy and Nathan would have had plenty of options.

Charging stations near Rehoboth, Delaware.

charging stations (*top*), liberally sprinkled with thousands of red dots, can conceal local shortcomings (*above*). Despite a seemingly liberal sprinkling of charging stations in the Rehoboth area, there were only three stations of the type they were seeking—and only one, miles from the beach, that could charge their car fast enough to salvage precious beach time. The result: an unanticipated change in plans and a lengthy delay. They ultimately settled on a slow charge closer to the beach. Their experience transforms an abstract array of dots into an anxiety-provoking treasure hunt. The story is told via a choreographed map and accompanying text; no portrait photos were used in order to protect Dorothy's and Nathan's privacy. Our "heroes" might better be described as victims in this case, but the point is that their travails personalized the story, allowed us to empathize, and turned a seemingly abstract topic into a relatable human drama.

The heroes you choose to profile don't necessarily have to be human. Some of the most powerful multimedia stories we've seen profile the adventures and life histories of mammals and birds, many utilizing tracking data from animals outfitted with GPS-enabled radio devices.

Florida's Archbold Biological Station and other partners produced a story called **Bear Necessities** ⊝, describing the adventures of M34, a young adult male black bear who was outfitted with a collar in Sebring, Florida. The bear wandered his home range for seven months until he suddenly took off, traveling over 500 miles in under two months. He crossed roads, skirted shopping centers, swam across lakes, and found shelter in wetlands and small woodlots. Animated maps track his progress, and video clips by Joseph Guthrie, a member of the capture team, describe his peregrinations.

Ultimately, he encountered Interstate 4, which bisects the state from Tampa to Daytona. Finally unable to conquer this daunting barrier, M34 returned to within 30 miles of his starting point. Meanwhile he had earned the admiration of his capture team: "M34 demonstrated astounding mobility and instinct for survival, all the while providing evidence of the landscape's fragile connectedness." The story ends with an appeal to support the Florida Wildlife Corridor, a plan to link a network of protected areas that was formally recognized in 2021—perhaps in part due to M34's dramatic story.

The advent of ever smaller and more powerful tracking devices is improving our understanding of animal movements, revealing countless stories of ultra-long aerial flights and terrestrial migrations that are often threatened by human activities. Tracking data presents huge storytelling potential; profiling migrating animals can personalize conservation, reminding us that we share this planet with animals whose dramatic lives deserve our attention, respect, and care.

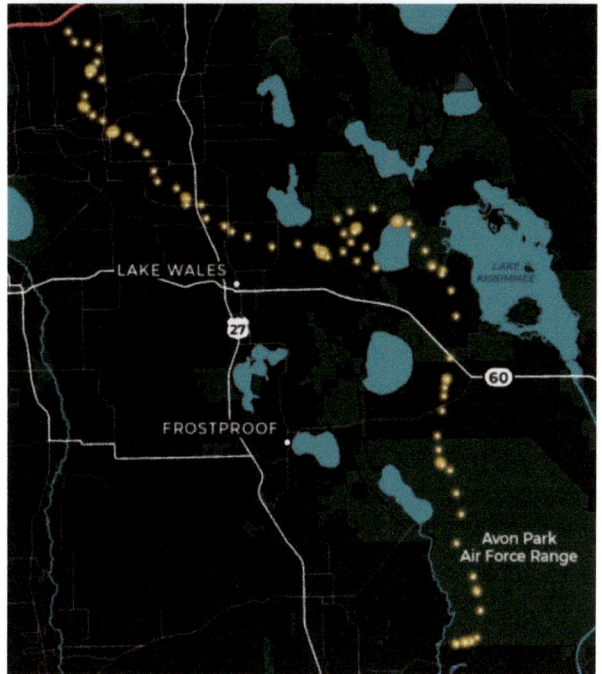

*Top*: A capture team measures Bear M34 and outfits him with a radio collar.

*Above*: A series of animated maps traces M34's travels.

## 3. Give your story rhythm

Establishing a repeating pattern isn't appropriate for all stories, but incorporating a rhythmic series of elements can lend structure to your narrative. Rhythm is comforting. Repetition enables your readers to settle in and anticipate a series of new items and new insights.

One of several rewarding collaborations with National Geographic Explorer Paul Salopek and our team resulted in an ArcGIS StoryMaps story called **Walking China's Antique Roads** ⊖. Salopek trekked the length of China as part of his "Out of Eden Walk" project, profiled in chapter 6, tracing the human diaspora from Africa to Patagonia. His China walk largely followed the paths of historical roads, ranging from the Burma Road's tortuous World War II–era supply route over the Himalayas to the eastern

Paul Salopek's *Walking China's Ancient Roads* story.

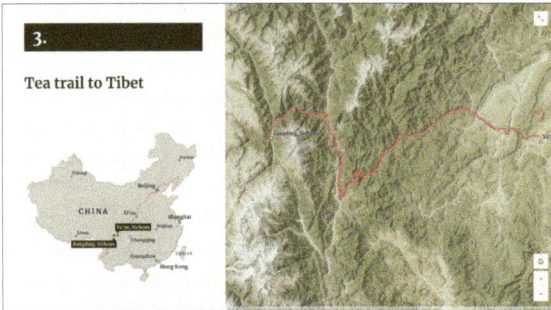

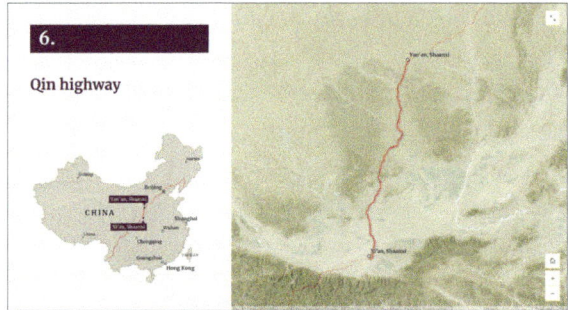

**22.8 billion chickens**

Chickens are far and away the most numerous type of livestock on the planet. There are about 155 chickens for every cow—and three for every human.

| | |
|---|---|
| China | 4.89 |
| Indonesia | 2.16 |
| United States | 1.97 |
| Brazil | 1.43 |
| Iran | 1.00 |

Chickens raised by country (in billions)

Wild chickens are believed to have originated in northern China, and were eventually domesticated in Southeast

**1.5 billion cattle**

Cattle are the second most common livestock animal. Their domestication is thought to have occurred roughly 10,500 years ago, in what is now considered the Middle East.

| | |
|---|---|
| Brazil | 214.9 |
| India | 185.? |
| United States | 93.7 |
| China | 83.21 |
| Ethiopia | 60.90 |

Cattle raised by country (in millions)

Today, these animals are especially prevalent in South America, where they're primarily raised for meat, and in India, where the animals are conserved

**1.2 billion sheep**

Sheep are believed to be one of the first domesticated animals, and are common throughout the Old World. They're especially prevalent in northeastern China, Central Asia, and North Africa, but are also raised intensively in New Zealand and Australia.

| | |
|---|---|
| China | 161.35 |
| Australia | 72.12 |
| India | 63.07 |
| Nigeria | 43.5 |
| Sudan | 40.57 |

Sheep raised by country (in millions)

Although New Zealand is famous for having more resident sheep than people

**1 billion goats**

These charismatic critters are the fifth most common kind of livestock on the planet. Goats were likely domesticated more than 10,000 years ago in modern-day Iran, and there are now more than 300 distinct species roaming the planet.

| | |
|---|---|
| China | 138.?? |
| India | 133.26 |
| Nigeria | 79.04 |
| Pakistan | 72.2 |
| Bangladesh | 59.71 |

Goats raised by country (in millions)

Maps, graphs, and artwork profile domesticated animal distribution.

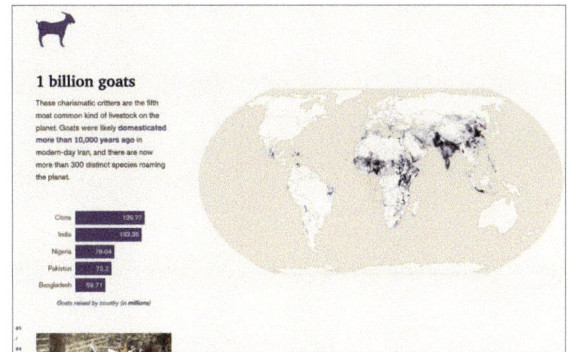

stretches of the Silk Road's network of Asia-spanning pathways. His story has a repeating drumbeat: Each section opens with a locator and regional map. Next comes a narrative illustrated with photos and describing his experiences; finally, he includes a "History" sidebar with additional maps and brief descriptions of the empires and dynasties with which his route intersected. The richly textured story paints a vivid picture of China's long history and its headlong rush into the 21st century.

Rhythm can also come with more concisely repeated multimedia treatments. Our story, **(Farm) Animal Planet** ↩, marries a series of maps showing the global distribution of domesticated animals with artwork, text, and graphs. Color treatments and silhouetted profiles differentiate the series, as do the dramatic differences, revealed within the maps, in the animals' distribution across the continents.

## 4. Create a little world

By definition, multimedia storytelling integrates multiple elements, often including maps, text, images, video, audio, embeds, quotes, buttons, links, and separators. Combining these elements can result in an engaging narrative. It can also result in a cacophony of visually conflicting elements.

A successful story will adopt a consistent visual and editorial style that immerses your readers in a seamless experience. That means not just writing an engaging piece of text; it means striving to unify all the elements of your narrative into a harmonious visual whole. One of the most effective ways to do this is to use color consistently and judiciously. A good approach is to choose a limited color palette and then deploy it throughout your story—in type treatments, maps, infographics, images, and other elements, such as separators and buttons.

**Of All the Fish in the Sea** ⊖ (*facing page*) by master storyteller Aaron Koelker combines science, history, and conservation messages into a seamless narrative that's simple in its visual approach but stunningly beautiful in appearance. All the maps and images are in black and white, with bright-red accents sparingly applied to maps, art, and graphics throughout the story. The result is a spare but immersive experience. The story has loads of information, but it's presented so simply that readers never feel lost or overwhelmed.

Prague's Institute of Planning and Development applied a very different treatment to its exploration of **The Diverse Prague** ⊖ (*facing page*), using bright colors in a variety of hues against a black background to evoke a sophisticated, urbane look and feel.

The ArcGIS StoryMaps theme builder assists authors in achieving a consistent, little-world effect. Our team has created six themes—Summit, Obsidian, Ridgeline, Mesa, Tidal, and Slate—that enable you to achieve a unified effect with a single click. Better yet, you can modify these themes or start from scratch, choosing from hundreds of fonts and a near-infinite range of colors to create a visual setting for your story. Your custom theme can match your organization's branding or, like a stage set, create an environment that increases your audience's engagement and amplifies your message.

| Theme | |
|---|---|
| **Summit** | Obsidian |
| **Ridgeline** | Mesa |
| **Tidal** | Slate |

Six default themes help give your story a distinctive style.

## 5. One size doesn't fit all

As I've mentioned elsewhere in this book, a challenge for designers who used to work solely in the analog realm is that one of the few things you could count on—the size of the publication you were designing for—is no longer static. We can view web-based multimedia stories on large screens, tablets, and other mobile devices, and in both portrait and landscape modes. There's a risk that what looks great on a big screen is cropped or otherwise compromised on smaller displays.

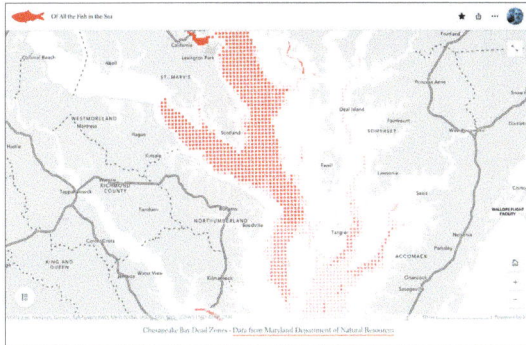

Little worlds: *Of All the Fish in the Sea.*

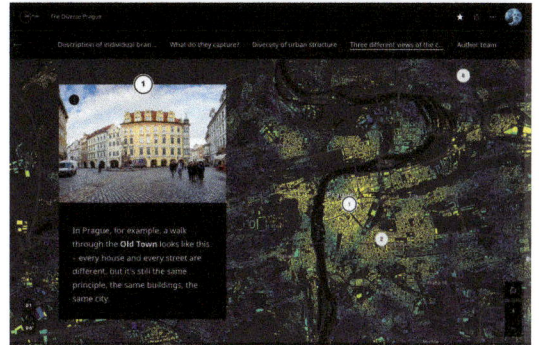

Little worlds: *The Diverse Prague.*

We've worked hard to make sure ArcGIS StoryMaps stories work equally well on a variety of screen sizes. But that doesn't mean you shouldn't actively anticipate how your narrative will look in these various contexts.

As you're working on your story, frequently check on how it's functioning in various screen sizes and proportions. The ArcGIS StoryMaps builder makes this easy: Click Preview in the header bar for a quick but accurate look at how your story will look on desktop, tablet, and mobile and in landscape and portrait orientations.

An example of a potential big screen, small screen pitfall is the use of full-screen photos in an immersive section. The effect looks great on large screens (1), but when viewed in portrait orientation on a mobile device (2), the focal point of the image might be cropped out. The solution: In the Image options panel (3) within the builder, move the white and gray focal-point indicator to the upper-right quadrant of the image. That way, the road will appear as part of the image regardless of screen size and orientation (4).

Telling stories on screens large and small.

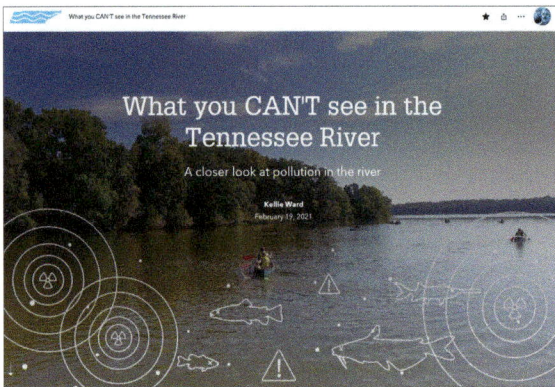

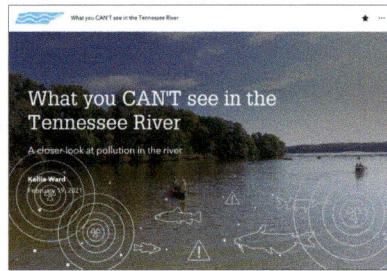

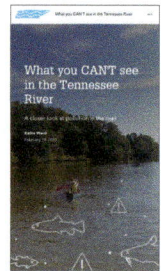

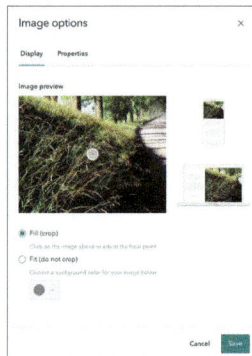

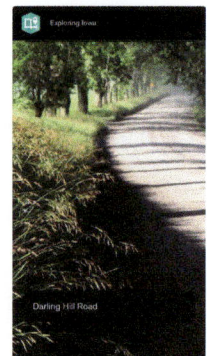

1

2

3

4

# 6. Think big, think small

If you're into maps, you know that scale is really, really important. A small-scale map provides an overview; a large-scale map shows local details. Similarly, changing scales can be an effective storytelling technique. Starting with a local anecdote or case study, then zooming out for the bigger picture, can be a good way to introduce a topic or issue and draw your readers into your story. Conversely, you can start with an overview to establish a broad context, then come in close to explore an issue more thoroughly and provide local examples.

Another way to think about this technique is **macrocosm versus microcosm**, with the latter defined by the *Oxford English Dictionary* as "a community, place, or situation regarded as encapsulating in miniature the characteristic qualities or features of something much larger."

For example, the Oaxaca hummingbird lives in a small area in Southern Mexico, an important area for biodiversity.

Estimated range of the Oaxaca hummingbird, *Eupherusa cyanophrys*, Map of Life (2019).

Like all hummingbirds, the tiny Oaxaca hummingbird plays a critical role as a pollinator, feeding on nectar and pollinating particular plants as it zips

The Half-Earth Project summarizes its mission in a narrative that moves from one species to many, and then to the whole world.

**Protecting enough biodiversity to ensure a healthy planet for future generations will require conserving half the planet for nature.**

Currently, about **15%** of Earth's lands and **7.5%** of its seas are under some form of protection.

While we are currently far short of the Half-Earth goal, progress toward 50% is being made daily.

50%          50%

14.9%        7.47%

**14.9%**
CURRENT LAND
PROTECTED AREAS

06
/
06

A multimedia narrative (*previous page*) by the Half-Earth Project, an initiative working toward protecting half our planet's land and water surface, begins with the tiny Oaxaca hummingbird, whose range is confined to southern Mexico. It represents a microcosm through its diminutive size and modest geographic extent and as a metaphor for a larger issue. The story describes the bird's important role in its ecosystem as a pollinator; then the story expands its focus, explaining how thousands of species play similarly key roles in the network of life. The narrative's focus expands farther, to a global scale, showing how current protected lands and waters fall short of the organization's ambitious goal.

Micro-to-macro and macro-to-micro can be equally effective techniques. So can a repeated pattern of overview and case study. Bring this oscillation to life with map choreography, and you'll likely have the makings of a powerful narrative.

## 7. Use active and passive maps

We've all used interactive maps, in which you can search, pan, zoom, tap, click, and swipe for a variety of purposes. Those of us who are geospatial professionals have a special fondness for map-based interactivity, without which our day-to-day work would likely be significantly hampered. As storytellers, we also love interactivity. We've come to realize, though, that interactive maps aren't always essential and that, in fact, they can detract from a narrative.

Creating a custom static map lets you present to your readers exactly what you want them to see and understand, without risk of distraction. As Esri cartographer Mark Harrower writes, "If zooming in on the map adds nothing to my understanding of the data, I feel annoyed. Put another way, if you can fit all of your data on the map and it still makes sense, you might not need interactivity."

Requiring readers to pan, zoom, and click on maps to spawn pop-ups is, in many cases, an imposition that interrupts the story's narrative flow. Several years ago, *The New York Times*, long a leader in producing elegant maps, infographics, and multimedia news stories, conducted research that showed that 85 percent of its readers don't bother to interact with its maps. Admittedly, the *Times*'s readership is far broader than ours at Esri. Because a higher proportion of our users are *cartophiles* who work with maps every day, we indulge ourselves—and our readers—with a higher proportion of interactive maps within the stories we produce. Often, we'll compromise, providing maps that support the narrative without requiring readers to

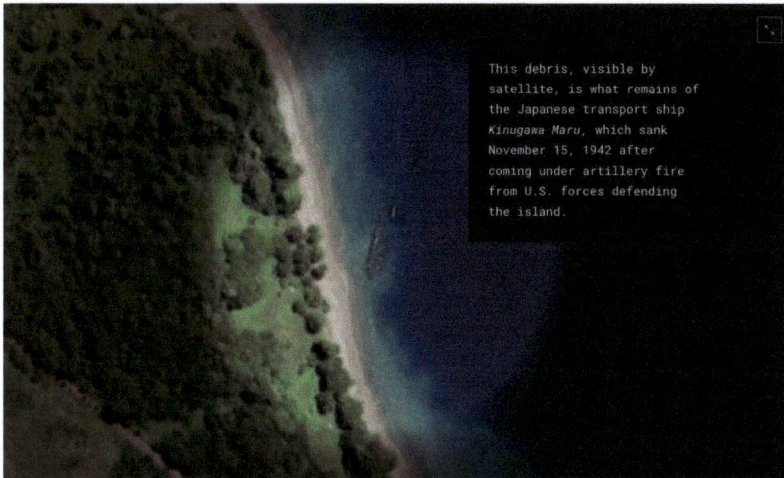

This debris, visible by satellite, is what remains of the Japanese transport ship *Kinugawa Maru*, which sank November 15, 1942 after coming under artillery fire from U.S. forces defending the island.

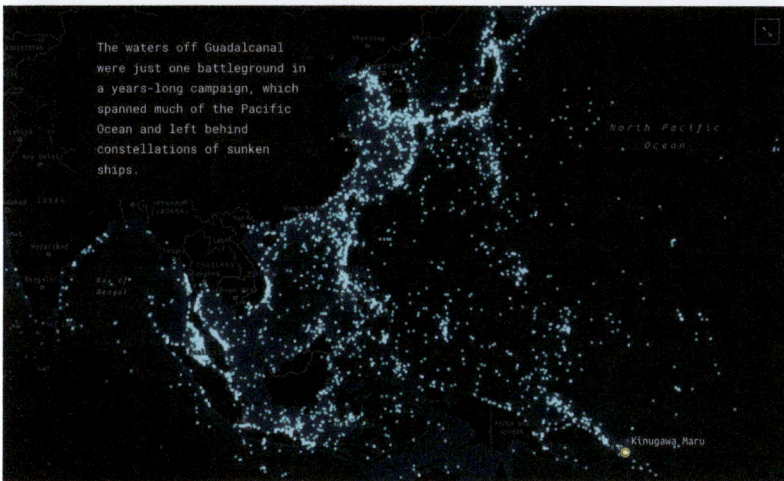

The waters off Guadalcanal were just one battleground in a years-long campaign, which spanned much of the Pacific Ocean and left behind constellations of sunken ships.

*Resurfacing the Past* begins with a single shipwreck (*top*), then transitions to a world map of more than 20,000 sinkings.

And the Pacific Ocean was just one theater of the war, which left virtually no corner of the global ocean untouched.

**8**

interact with them but that give readers the *option* to pan, zoom, and click. But it's important to remember that even that option comes at a cost: Interactive maps take longer to load than static maps that are incorporated into stories as images.

Our favorite technique is to make maps lively by *simulating* interactivity, primarily by incorporating map series into immersive sections—using the map choreography that I discuss in chapter 8. One of my favorite stories, **Resurfacing the Past** ⊖ (*previous page*) by my colleague Cooper Thomas, is a prime example of both simulated interactivity and the think big, think small principle we just discussed. His story starts locally, spotlighting a single World War II–era shipwreck, and then zooms out to a world map locating more than 20,000 shipwrecks. The map zooms in increments as readers scroll. No clicking required.

## 8. Keep it short and sweet

We know you love your work, or your organization, or your cause. And we know you can describe your passions in endless detail. But many of us don't have the patience to dive deep into a lengthy multimedia story—especially on the web, a medium that is rife with temptations to click away.

If your story is well told, if its images are beautiful, and if its subject matter is compelling, readers will likely stick around, at least for several minutes. Our analytics show that stories keep people engaged for an average of six minutes, which is far longer than the 30 seconds or so spent on a typical web page. Still, it's best to strive for brevity. If your story is nearing novel length, consider splitting it into a series of stories and featuring them in a collection. That's what I did for a story on **Climate Migrants** ⊖ (*left*). I had drafted it in several sections, each on a relatively discrete topic. I figured it was fine as a single story until a former colleague gently suggested that it would be more effective as a collection of individual stories. That gave me the opportunity to give each story its own cover treatment and to feature them on a collection page, which offers several attractive design options.

*Climate Migrants*: One long story vs. five shorter stories.

## 9. Make a call to action

Now that you've inspired your audience, don't leave your readers hanging. Give them something to do—even if it's just providing a link to more information. If you're telling a story about a cause or issue, it's doubly important to conclude your narrative with one or more calls to action. A successful story doesn't just inform readers, it inspires them. You want to provide opportunities for people to act on their inspiration.

A call to action might be as simple as adding a Donate button. Or it might include several options. Tompkins Conservation's excellent story on rewilding, **From Dying to Wild Again** ⟨⟩ (*right*), begins with a touching anecdote about a stranded jaguar (a fine example of featuring a nonhuman hero as a storytelling device) and ends with links to information resources, a sound clip of a growling jaguar (the hero returns), a Subscribe and Donate button, and a card linking to a TED Talks presentation.

A caution: Giving readers too many action options risks reducing the response. Faced with too many choices, some of us will probably just shrug and leave the site.

We hope you'll find these nine tips helpful. We've realized there's a not insignificant learning curve involved in mastering the many mechanics of building a story. But the greater challenge, and the more exciting one, is in creating a narrative that captures readers' hearts and inspires them to action. Storytelling is an art. As with all artistic endeavors, success rarely comes without careful planning and repeated polishing.

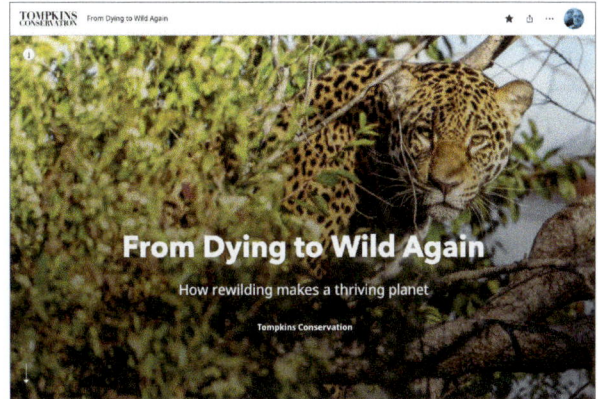

Tompkins Conservation makes the case for rewilding and points readers toward action options.

# Divots and Dashboards

## Golfing with a data-driven caddy

**Author:** Warren Davison, ArcGIS StoryMaps team

**Medium:** ArcGIS StoryMaps

**Story behind the story:**
"Using geographic information systems (GIS), spatial analysis can be applied to nearly any phenomenon, spanning the globe or 18 holes of golf," wrote ArcGIS StoryMaps team member Warren Davison. He proceeded, courageously, to prove his point by analyzing his golf game.

**Why it's special:** Warren used a mobile app to record the location and club choice for every shot during several 18-hole rounds and used the location data to determine each stroke's distance and direction. He mapped each round, memorializing his triumphs and tragedies.

**The author:** As part of his analysis, Warren aggregated all his tee shots (*facing page, top*), revealing that his drives, on average, tended to drift rightward. Then he used his considerable design and illustration talents to package and present his analysis as an entertaining and engaging story.

*Divots and dashboards*

Golfing with a data-driven caddy

**Warren Davison, Esri's StoryMaps team**
November 1, 2023

**Divots and Dashboards: Scores**
Improving. One swing at a time.

| Player | 1 | 2 | 3 | 4 | 5 | 6 | 7 | 8 | 9 | OUT | 10 | 11 | 12 | 13 | 14 | 15 | 16 | 17 | 18 | IN | Total |
|--------|---|---|---|---|---|---|---|---|---|-----|----|----|----|----|----|----|----|----|----|----|-------|
| Par | 4 | 4 | 4 | 4 | 4 | 3 | 4 | 3 | 4 | 32 | 3 | 4 | 4 | 3 | 4 | 4 | 4 | 3 | 4 | 30 | 62 |
| War… | 4 | 3 | 6 | 4 | 4 | 3 | 6 | 3 | 5 | 38 | 3 | 4 | 5 | 5 | 4 | 4 | 5 | 4 | 40 | 78 |
| Ryan | 4 | 5 | 5 | 5 | 4 | 4 | 5 | 4 | 6 | 42 | 4 | 5 | 7 | 5 | 4 | 4 | 4 | 4 | 43 | 85 |

*Top*: Cover art and title.

*Above*: A round of golf, mapped and quantified.

*Left*: One of several illustrations created by Warren for his story.

*Top*: Warren aggregated his tee shots into a single imaginary hole in order to discern overall patterns.

*Above*: Warren's map of every shot made during three rounds of golf at Rockway Golf Course.

# Hot Numbers

It's a fact: Human impacts are warming our climate and changing our planet

**Author:** Allen Carroll

**Medium:** ArcGIS StoryMaps

**Story behind the story:** Even as the evidence of human-induced climate change became nearly impossible to refute, a couple of people in my family continued to be climate skeptics. That prompted me to use our storytelling platform to present the evidence and—ideally—convince them, and others like them, that climate change is real.

**Why it's special:** I chose a dozen key statistics, which I customized as images and combined with graphics to provide rhythm and visual consistency and to enliven what could have been a dry narrative. As readers scroll, each number panel is followed by brief interpretive text, maps, charts, and links, as the sea ice example (*facing page, left*) shows.

A concluding "Action" section provides resources for speaking out, calculating your own carbon foot-print, offsetting your emissions, and supporting organizations working to combat climate change.

**The author:** Did I change any minds? Did I convince my relatives? I confess to having failed, in the interest of family harmony, to present this story to them. Shame on me.

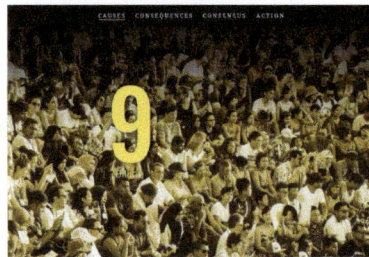

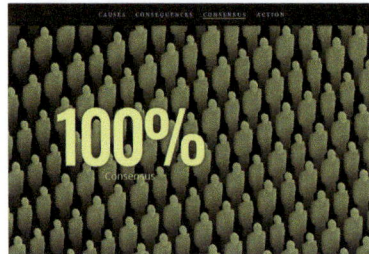

Cover title (*top*) and four of 12 section openers (*bottom four*) presenting key climate-related statistics.

CAUSES   CONSEQUENCES   CONSENSUS   ACTION

# 13%
Reduction

**Every 10 years the extent of Arctic sea ice in September declines by 12.8 percent.**

Arctic sea ice reaches its minimum extent each September. That extent is declining at a rate of 12.8 percent per decade relative to the 1981-2010 average.

The graph above shows the average monthly Arctic sea ice extent each September since 1979, derived from satellite observations. The 2012 extent is the lowest yet recorded. The animation below is based on satellite observations and shows minimum Arctic ice extent since 1979.

Source: NASA

1979 ———————————○———————————— 2018

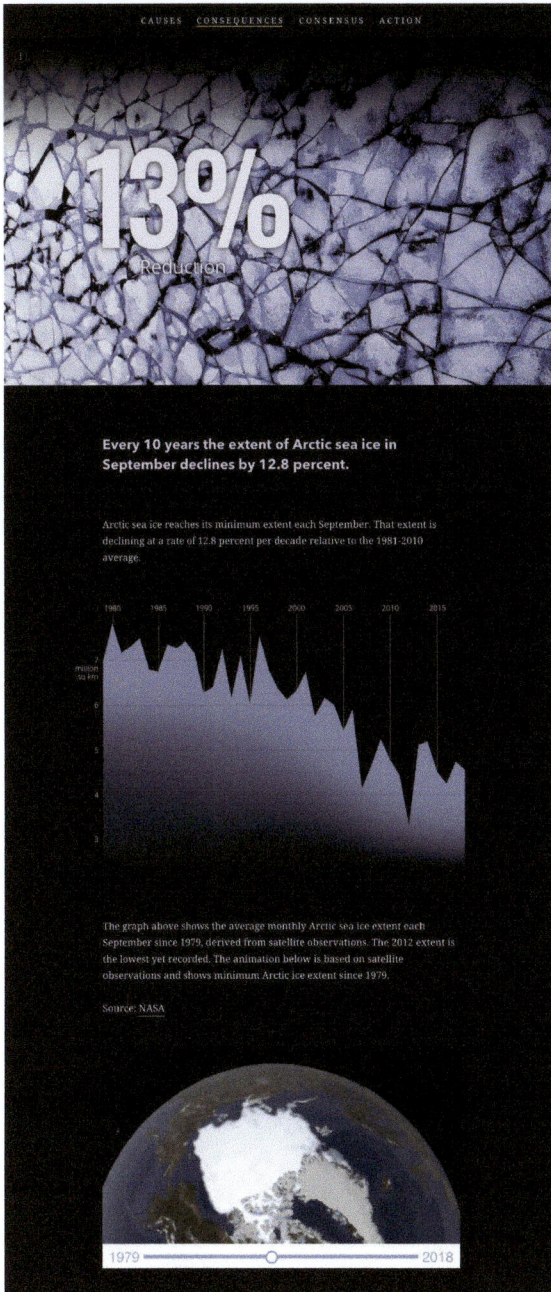

Each statistic is reinforced by maps, infographics, and links to authoritative sources.

CAUSES   CONSEQUENCES   CONSENSUS   ACTION

# ACTION

**We all need to take action.**

Climate change isn't the fault of some distant villain. All of us contribute to the greenhouse gas emissions that are changing Earth's climate. And all of us can help alleviate the problem. Here are a few things that you can do.

### Speak out

As climate scientist Katharine Hayhoe says, "Talk about it. Advocate for change at every level from your family to your school or business or organization to your elected representatives." As much as we can work as individuals to reduce our climate impact, governments and large corporations have a far greater effect on climate—and will be much more likely to make reforms if we put pressure on them.

[ Katharine Hayhoe's TED talk ]

### Calculate your carbon footprint

A key first step in helping reduce your impact on the Earth is to assess the size of your impact. Several organizations offer lists or, better yet, step-by-step online questionnaires that make it easy to estimate your footprint. One of the better ones is the University of California's CoolClimate Calculator.

[ Calculate your footprint ]

### Reduce your carbon footprint

A number of reputable non-profits provide lists of steps you can take to reduce your personal and household footprint. Global Stewards offers a list of Top 20 Ways to Reduce Your Carbon Footprint.

[ Reduce your footprint ]

### Offset your carbon emissions

One way to alleviate your impact is to contribute to efforts, such as renewable energy projects, that are reducing overall levels of carbon emissions. This article by the Natural Resources Defense Council discusses the pros and cons of carbon offsets. Conserve Energy Future lists 11 Best and Popular U.S. Offset Providers.

[ 11 offset providers ]

### Donate, join, or volunteer to organizations

Hundreds of organizations are mobilizing to tackle the many issues related to climate change and sustainability. Project Drawdown gathers and facilitates a broad coalition of researchers, scientists, policy makers, business leaders and activists to assemble and present the best available information on climate solutions. 350.org seeks to reduce atmospheric CO2 levels. The Nature Conservancy protects biodiversity by preserving key lands and waters. World Wildlife Fund and the Wildlife Conservation Society work to protect wildlife in a changing world.

[ Project Drawdown ]

Several means by which readers can take action and learn more about climate change.

# 8

## Plan, Produce, Polish, Publish

A digital spyglass peers through modern-day Manhattan to hint at the city as it was in 1836.

# Creating ArcGIS StoryMaps stories

ArcGIS StoryMaps is Esri's storytelling resource. It enables the creation of multimedia narratives combining interactive maps with text, photos, videos, audio, and embedded content. More than three million ArcGIS StoryMaps stories are hosted in ArcGIS Online, Esri's cloud service. They're powerful and versatile, with features that enable many forms of storytelling, from simple reports created by elementary school students to technical narratives serving specialized audiences.

This chapter guides you through the process of conceptualizing, building, publishing, and maintaining ArcGIS StoryMaps stories. I've organized the process into four alliterative steps: Plan, Produce, Polish, Publish.

We're about to take a rather deep dive into the story creation process, but please don't be intimidated by the length of this chapter. Rest assured that creating ArcGIS StoryMaps stories is intuitive and fun. The ArcGIS Story-Maps team has invested years of effort into democratizing the storytelling process, making sophisticated functions available without requiring story authors to have web development or programming skills.

Although this chapter will cover the features and functions of ArcGIS StoryMaps itself, it will also touch on the creative process of telling stories. ArcGIS StoryMaps is not unlike word processing software. You can use Microsoft Word to create a two-paragraph cover letter or to write the Great American Novel. Similarly, you might use ArcGIS StoryMaps to dash off a walking tour in an afternoon or work for weeks—even months—on a collection of stories that provides inspiration and insights to influential audiences. The creative process is messy, iterative, sometimes difficult, and often deeply rewarding. I hope the steps outlined here will help you discover and reap the rewards of effective storytelling.

# Step 1: Plan

## What's your message?

It's best not to dive right into the ArcGIS StoryMaps builder. Rather, you should spend some time articulating the **main message** of your proposed narrative. Ideally, you should distill your intentions into a single sentence that defines your goal. An example: My team has produced a series of stories called **Living in the Age of Humans** ⊝, including installments on the human reach, living land, forests, and biodiversity. I proposed that, after a lengthy hiatus, we revive the series with a story on the ocean. I defined the main concept, rather prosaically, as "a broad overview of human impacts on the ocean, starting with a series of beautiful images and including maps from ArcGIS Living Atlas of the World and other sources."

If a member of our team has a story idea, we ask them to fill out a pitch template that we use to articulate story concepts and make the case for investing time and effort in a new project. The team leads and I then evaluate the pitches before we green-light or decline story concepts. A somewhat abridged version of the pitch I prepared for the oceans story is shown below.

The pitch template helps us ensure that the stories we tackle make strategic sense and align with our goals. Approved pitches go into a shared

### Pitch Template: Oceans

**What and Why?**

| What is the purpose of your story? Why are you making it? What question(s) does it solve? | Our Anthropocene series has been perennially popular. Story would provide a broad overview of human impacts on the ocean, starting with a series of beautiful images and including maps from ArcGIS Living Atlas and other sources. |
|---|---|

**Value**

| Where does this idea fit in our strategic plan? Who is the intended audience? | • Aligns with company ethos<br>• Expands on already popular series<br>• Primary audience: students and educators |
|---|---|

**Scope**

| What will be included? What is not included? What content/data and permissions do you already have, and what is needed? | Potential ArcGIS Living Atlas content:<br>• Global ocean conservation priorities<br>• Ocean Health Index<br>• Marine protected areas<br>• Sea surface temp, acidification<br>Not in scope: Physical oceanography |
|---|---|

**Potential Stakeholders**

| Who may be interested in this project? | Esri Ocean, Weather & Climate GIS Forum |
|---|---|

**Proposed Timeline**

| Can this project be completed during any sprint or is it time sensitive? Does this need to be promoted on a specific day, i.e., a holiday or event? | Publish on World Ocean Day, June 8? Or put off until next ocean forum |
|---|---|

**Key Takeaways for Readers**

| What will readers know after they finish your story? What do you want someone to do, think, or feel after they've finished reading? Provide 1-3 takeaways. | Raise awareness about human impacts on the ocean:<br>• Climate change<br>• Fisheries are being overexploited<br>• Growth of ocean conservation |
|---|---|

**Level of Effort**

| How long with this project take? (A sprint, a milestone, etc.) How many people will it require? | High level of effort. 3-4 sprints, 3-4 team members involved |
|---|---|

pitch tracker table that lists team participants and helps us track progress toward completion.

## Who is your audience?

No pitch should be made, no concept finalized, without a strong notion of who its intended audience is. A story about grizzly bears that's targeted toward fifth graders will be very different from one aimed at conservation biologists. We strive to make our own stories as broadly appealing as possible, but many of our efforts are aimed at GIS professionals. We can be confident that GIS pros will know the differences among web maps, layers, feature services, and so on. But for broader audiences, we're careful to avoid terms that might confuse readers.

It can be helpful to have a persona in mind as you develop a story. You can imagine an individual that typifies the audience you're trying to reach. Keeping that person (or audience) in mind can help make your storytelling more personal and less preachy or academic than one aimed at a class or colleague. You might consider aiming your story at a specific subaudience. Perhaps your organization is hoping to collaborate more closely with a partner. Maybe you're doing a project assigned to you by a manager or teacher whom you want to impress.

Hannah Wilber, a former member of the ArcGIS StoryMaps team, wrote in a perennially popular blog post, "Understanding your audience, then tailoring your story to align with their interests and knowledge, is perhaps the most foundational part of creating an effective story."

> ❝
> Understanding your audience, then tailoring your story to align with their interests and knowledge, is perhaps the most foundational part of creating an effective story.
>
> **Hannah Wilber**
> Formerly of the ArcGIS StoryMaps team at Esri

## What's the hook?

It's best to avoid gimmicks, but it's perfectly fine to actively look for clever ways to draw attention to your story. Can you launch or tout your story at a specific conference or event? Can you help attract an audience by tying your narrative to an upcoming holiday or anniversary? Can you use words or catchphrases in your title and subtitle to make your story more visible to search engines? It's good to consider, even at this early planning stage, how your story will find an audience. We're all beneficiaries—and victims—of an overwhelming deluge of content. For story authors, the challenge is to make narratives stand out, in one way or another, amid this deluge.

## What about maps?

We've encountered plenty of ArcGIS StoryMaps stories that don't use a single map. Many of those stories are highly effective. But from the start, we've envisioned our platform as a means by which stories can benefit from the depth, context, and insights that geography can bring to most topics. As you plan your story, think hard about the roles that maps might play in your narrative, how they can enrich your story, and how maps, visuals, audio, and text can be woven together into an engaging and impactful experience.

## Look for inspiration

Now that you have a concept and plan, it's almost time to develop a story outline. I suggest that you first take a brief intermediate step to seek inspiration from exemplary stories. The ArcGIS StoryMaps gallery, accessible from the header bar on our website, presents scores of outstanding narratives representing a variety of topics and approaches. The gallery is searchable, and you can filter it by content type, features, industry, topics, and use cases.

The gallery presents a tiny subset of the millions of hosted stories. You can find many more through the ArcGIS Online search function or by typing a keyword or keywords and the word "StoryMap" in a search engine.

The ArcGIS StoryMaps gallery and its filter function.

## Define your key takeaways

It's likely that you're excited about the topic of your story and are eager to share your knowledge. That's great, but it can lead to a pitfall, which is to overload your story with details and related material that's not essential to your narrative. It's important from the start to clearly define the message you want to convey and identify the items you want your audience to take with them after reading your story. It's best to limit your takeaways to three at most; otherwise, you may confuse or overwhelm your readers with too much information.

## Begin a content inventory

Drafting a list of proposed content for your story is a good way to organize your thoughts. The list can include editorial concepts, anecdotes, names of individuals, and data sources, as well as maps, infographics, photos, videos, sound files, and embeds. You can tag the items to which you don't have ready access and add them to a to-do list. Your content list will no doubt change as you refine your storyboard and build your narrative.

## Outline your story

A common way to outline a story, article, or paper is to use Microsoft Word or PowerPoint, typing story elements as a numbered series of entries or a list of bullet points. It's an acceptable option, but there's a risk that words will take precedence over maps and visuals. It's all too easy to develop a narrative as text and then consider, almost as an afterthought, how that text can be supported by visuals and maps. It's important, as you develop a story outline, to imagine all the elements as equal players. That's why I prefer sketching, either on paper or, better yet, on a whiteboard.

With a whiteboard you—or, ideally, you and a couple of colleagues—can draw boxes, arrows, and keywords, erasing and redoing as you go. Adding movable sticky notes to your whiteboard can give you maximum flexibility to move story elements around.

When you're done, simply take snapshots of your sketches with your phone or transcribe them into text or PowerPoint. It's advisable to do this quickly: scribbles that were dense with meaning during a creative session

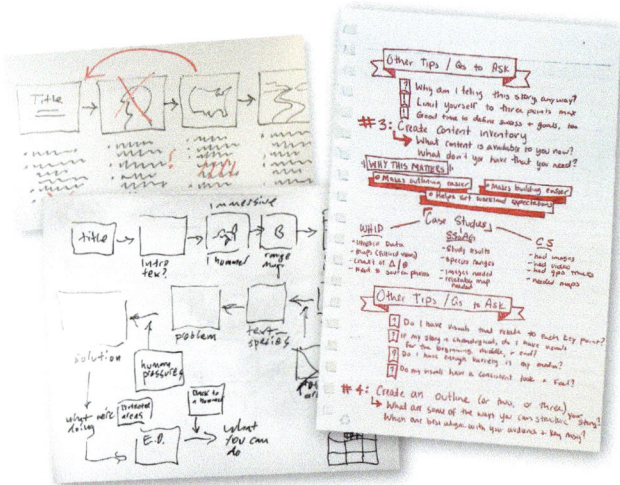

Storyboards.

might be indecipherable a day or two later, when you may have forgotten what you meant. Finally, don't treat your storyboard as authoritative. Creating a multimedia story is a messy, iterative process. Your final story may bear little resemblance to your initial outline. Don't be tied to your initial ideas. Be willing to give up your cherished favorites if it becomes apparent that doing so makes for a stronger narrative.

## Refine your ideas

A first-draft storyboard might elegantly accommodate the key takeaways and supporting information for your planned story. But it might also read like the outline for a school report, presenting the facts efficiently but conventionally.

I suggest you sleep on your storyboard. Not literally—that would be uncomfortable—but take a fresh look at your outline after a day or two and think carefully about how you can make your story more engaging. Does your storyline have an arc of some sort, taking the reader on a journey that ends in a moral or resolution? Is there a hero or character that can personify or personalize your story? Can you find a way to make your readers emotionally connect with your story?

Many professionals—especially scientists—are trained to be objective and dispassionate, removing personal anecdotes and viewpoints from their writing. But the elements that are forbidden in a scientific paper might be the very things that can make an ordinary story extraordinary and provide the sort of human connection that can evoke your readers' empathy.

Dr. Rae Wynn-Grant, a large-carnivore ecologist, created a story about her work with grizzly bears in Montana's American Prairie reserve. She introduced herself within the **Homecoming** ⟶ story and recounted her fieldwork in the first person. The result is an engaging narrative that reveals a young scientist's passion for nature. This is one of many techniques that can turn an ordinary tale into a memorable experience.

# Homecoming

How grizzly bears can find safe passage to their home ranges in the American Prairie Reserve

Dr. Rae Wynn-Grant

Grizzly Bear Conservation    American Prairie Reserve    Barriers and Opportunities    Moving Forward    About this story map

## My name is Dr. Rae Wynn-Grant, and I am a large carnivore ecologist.

One of my passions is exploring human-carnivore coexistence, and the return of grizzly bears to the ecosystem in and around the American Prairie Reserve (APR) is a great study system to do this work.

Historically, humans and grizzlies shared these prairie landscapes, but today the journey of these bears from protected areas in the West to the APR landscape in the East poses many unique challenges that they have not evolved to properly understand. I use a combination of field-based data collection, high level statistical tools, and Esri mapping technology to help these bears return home and help people better live with them.

Photo: Carolyn Barnwell

## Grizzly Bear Conservation

Grizzly bears are an iconic North American species that are recovering from near-extinction in many parts of the continent.

Listed as "threatened" under the US Endangered Species Act, long-term conservation action has allowed them to rebound in many areas, including parts of Montana.

Photo: Uryadnikov Sergey

Rae Wynn-Grant's first-person story about grizzly bears in Montana.

## Solo or ensemble?

I've created many ArcGIS StoryMaps stories by myself. Doing so can be rewarding, if a bit lonely. And one can enjoy pride of sole ownership. But I highly recommend that you collaborate with colleagues in building stories. Without exception, I've found that creating stories collaboratively results in stronger, more effective narratives. People with different backgrounds and predilections inevitably bring fresh ideas and approaches. If you're a GIS professional creating a story for an organization, I especially recommend that you work with a communications professional in crafting your story. When we meet with an organization to advise them on storytelling, we're more optimistic about the outcome when both GIS people and communications people are at the table.

Story collaborations can take many forms, from coauthorship to an occasional review and editing session. Collaborators can perform tasks associated with their specialties, including cartography, graphic design, data visualization, and wordsmithing. Collaboration also has its limits: Working with a team of more than four or five can lead to diminishing returns. Storytelling by committee is inefficient and is likely to rob a story of its personality and style.

## A caveat

The process I'm outlining is based on 14 years of creating multimedia stories—and countless editorial projects before that—which means that I've learned that every creative process has its own requirements, cast of characters, circumstances and, yes, politics. Your process may vary. Be flexible, be diplomatic, and have fun.

# Step 2: Produce

Let's say you have a storyboard or outline you're comfortable with. And you've assembled much of the story content. It's time to start building.

As much as I'd like to offer comprehensive, step-by-step instructions on these pages, it's just not practical to do so. ArcGIS StoryMaps is updated every couple of weeks, so a detailed description is bound to quickly become out-of-date.

The ArcGIS StoryMaps website, the ArcGIS Blog, and Esri Academy all have abundant resources to help you become an expert storyteller. But an instructional story called **Getting Started with ArcGIS StoryMaps** ⊖ is perhaps the best single source for new storytellers, and it's one that we

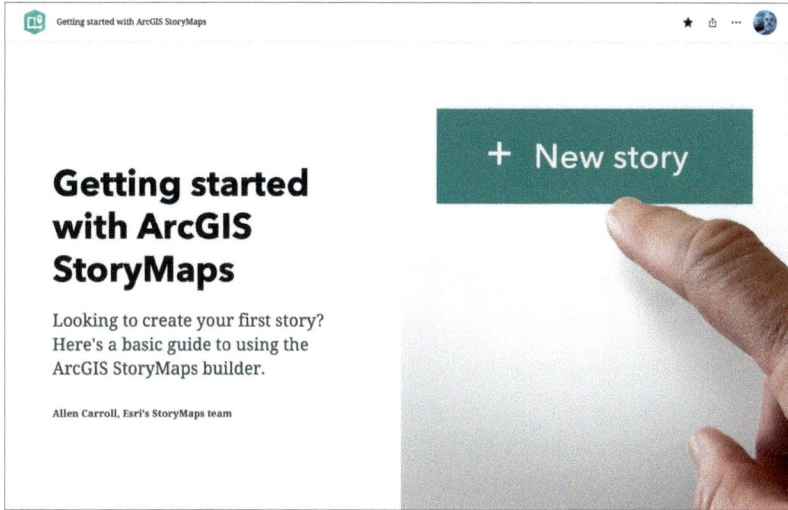

Getting started with ArcGIS StoryMaps.

conscientiously update with each new release. This section follows the sequence of items in the *Getting Started* story but offers some additional commentary, a few personal opinions, and some occasional advice. You might want to have your laptop open to the *Getting Started* story as you read this section (go to links.esri.com/StoryMapsTutorials and click the title).

## The building blocks

The *Getting Started* story will lead you to the **My Projects** page, where you'll be able to access your content. On the left side of the page, you'll see buttons accessing the ArcGIS StoryMaps four output types. **Stories** are the familiar scroll-driven multimedia narratives that now number in the millions. **Briefings** are a relatively new but popular slide-driven presentation format. Briefings have their own builder function that includes many of the functions you'll find in stories. **Collections** provide a convenient way to create pages that aggregate a series of stories and other content. **Themes** let you create distinctive graphic styles for your stories. We'll focus first on stories.

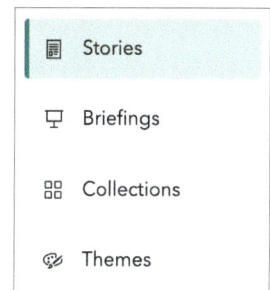

ArcGIS StoryMaps has four output types.

After you've created a draft cover with title, subtitle, byline, and image (we'll discuss cover options and refinements shortly), you'll scroll down and click Create New Story (a blue plus sign), which will reveal a menu of options. This menu, which we call the **block palette**, is the heart of the ArcGIS StoryMaps builder. It gives you a set of options—think of them as building blocks—with which you can assemble your story, piece by piece. The *Getting Started* story tells you what you need to know about the palette.

**Basic**

A≣ Text

▭ Button

-- Separator

{ } Code

▦ Table

**Media**

⌖ Map

🖾 Image

🖾 Image gallery

▶ Video

◁)) Audio

</> Embed

⊞ Swipe

🕐 Timeline

**Immersive**

▥ Sidecar

⚲₀ Map tour

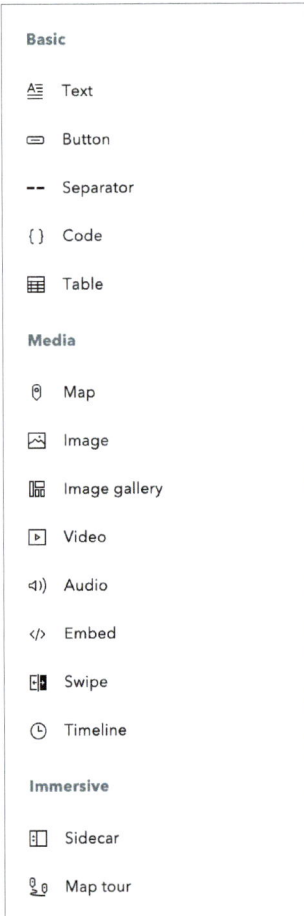

The block palette.

It's organized into three groups: **basic**, **media**, and **immersive**. Basic functions include text; media embraces maps, images, video, and so on; and immersive features two versatile blocks, sidecar and map tour.

Once you've added text to your story, you can highlight words or passages to access a second palette that lets you adjust text sizes and formats.

If you're a neophyte, it may be challenging to envision how these various building blocks—especially the immersive ones—can be used as narrative devices. That's why we maintain a gallery of exemplary stories. You can access it by clicking the Explore Stories item on the My Projects page. It provides dozens of inspiring examples, many of which employ a variety of blocks and type styles.

One of the basic blocks in the palette menu is **Separator**, which can be helpful in providing a visual break between parts of your story. Separators are simple graphic elements by default, and they serve to delineate a break or transition within a story. A convenient feature of ArcGIS StoryMaps is the ability to create custom separators. They can add style and flavor to stories, complementing and reinforcing your story's editorial theme. If care isn't taken, they can also become too dominant or too cute—or both.

Here are some examples from stories created by our team for stories on oceans, electric vehicle charging challenges, and on-the-ground conservation. You can incorporate custom separators within the theme builder.

Custom separators.

## Working with images

The ArcGIS StoryMaps builder gives you a lot of options for incorporating and refining images. You can add photos by clicking **Image** in the block palette. When you hover over an image within your draft story, you'll see a set of options in a little menu at the top of the image. The pencil icon gives you access to a set of editing tools with which you can crop, rotate, flip, and annotate your images. I often use Photoshop to adjust and crop photos, but many storytellers lack access to sophisticated desktop publishing software,

whose myriad functions may amount to overkill regardless. When crop-
ping, you can choose from a set of standard rectangular sizes, set your own
custom size, and even give the image rounded corners or an oval shape. You
can also rotate the image or flip it horizontally. On the markup tab, you'll
find drawing tools—lines, points, shapes, arrow, and so on—that you can
use to annotate your image to call out specific elements of it.

Image tools.

Annotated photo.

Size options for images include—as you might expect—small, medium,
and large. Additional float left and float right options let you inset small
images into your text. Experiment with size options as you build your story,
and, as you do, use the preview function to see how your photos will look
on tablets and mobile devices. Photo treatments will vary depending on
the type of story you're telling. For highly visual topics, don't be afraid
to vary the size of your images. An occasional full-width image can add
impact and variety to your story.

An **image gallery** function lets you display groups of photos and provides
three layout options (*right*): dynamic squares, jigsaw, and filmstrip.

The image gallery is ideal for displaying
a group of related topics or objects, such
as these mushrooms, photographed by
editorial team member Ross Donihue. I
advise against using a gallery to display
images that are individually integral to
your narrative or simply because you
like the photos and want to show them
off. Galleries can add visual clutter, and
asking readers to click through them can
interrupt the narrative flow. In addition,
most of your readers may simply skip
over the gallery. My advice: In most circumstances, it's better to feature a single,
memorable image rather than a cluster
of them.

Image gallery layout
options.

Dynamic squares

Jigsaw

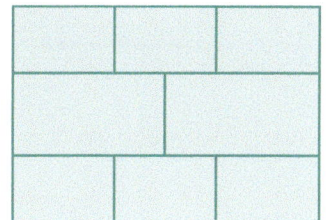

Filmstrip

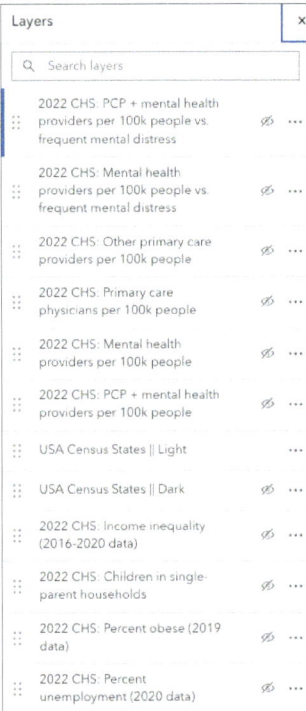

Part of the layers list for a web map supporting a story on America's mental health crisis.

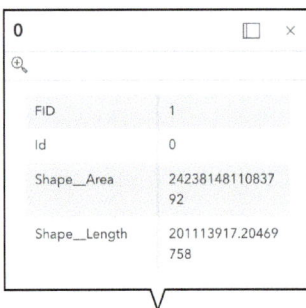

An unconfigured pop-up.

# Working with maps

ArcGIS StoryMaps features two mapping capabilities: **Express maps** allow you to easily create simple maps to show locations, routes, areas, and other basic features. We'll return to that a little later. The second capability taps the sprawling ecosystem of content and capabilities of Esri's cloud resource, ArcGIS Online. The myriad maps, layers, services, and other items in ArcGIS Online constitute what is probably the world's largest aggregation of accessible geographic data. But it should be used with caution. Some items may require permission to use; others might be created by individuals (students, for example) and aren't authoritative. The riches within ArcGIS Living Atlas of the World, a curated subset of ArcGIS Online, are authoritative, but even within ArcGIS Living Atlas, items may disappear over time or require permission to use. Much of the data in ArcGIS Online has been created by GIS professionals and may include obscure numerical fields and technical descriptions that may not be appropriate for your audience.

When our team is working on a story, we frequently use ArcGIS Online Map Viewer to create and refine a companion web map that includes the layers that we plan to feature in the story. We save it with all the layers turned off, except for the basemap. As we build the story, we can repeatedly incorporate the web map and use the map controls within the ArcGIS StoryMaps builder to expose appropriate layers. We work throughout this process to ensure that the cartography is clear and uncluttered and that the map design is consistent with the look and feel of the story in which it will appear.

A common pitfall of the use of web maps as an ingredient in the storytelling formula is **unconfigured pop-ups**. Clicking web map elements shows all their attributes by default; unless they're customized, these attributes create only clutter and confusion. ArcGIS Online Map Viewer provides ways to delete and modify pop-up elements or disable them; other tools, including a capability called ArcGIS Arcade expressions, enable additional refinements. We encounter plenty of stories that are otherwise beautifully designed and executed but whose pop-ups reveal obscure data tables such as the example at left. Remember: Configure or disable your pop-ups!

Now, back to express maps. This feature was designed to enable people with no GIS skills to create simple but beautiful maps. In fact, storytellers at all levels can find express maps useful. As Will Hackney of our ArcGIS StoryMaps team puts it, "Express maps, at their core, are intended to be clean, simple complements to storytelling, relating pertinent geographic context in an immediately digestible way."

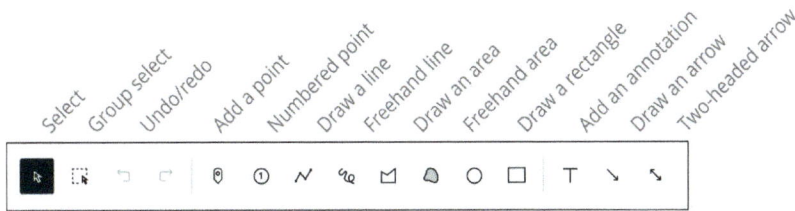

Select   Group select   Undo/redo   Add a point   Numbered point   Draw a line   Freehand line   Draw an area   Freehand area   Draw a rectangle   Add an annotation   Draw an arrow   Two-headed arrow

The express maps toolbar.

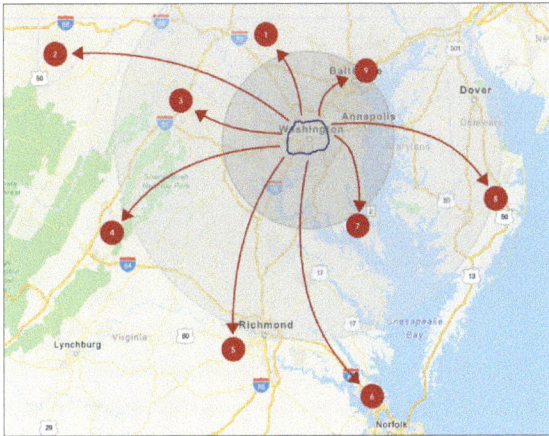

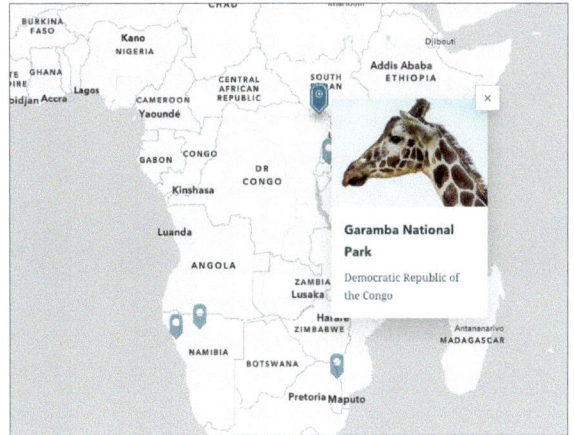

These tools can be used to build maps that perform myriad functions, from simple locators to maps that identify multiple features and show routes and movement. A couple of examples are shown above. Express maps can use a variety of basemap styles, and you can create pop-ups with images, titles, and descriptions.

Express maps locate weekend destinations near Washington, DC (*left*). Customized pop-ups identify African parks (*right*).

## Multimedia blocks

At the bottom of the block palette are four functions—sidecar, map tour, swipe, and timeline— that present all sorts of exciting creative possibilities. We call sidecar and map tour immersive because they take over the whole browser window as readers scroll. All are optional but add depth and variety to the reading experience.

## Sidecar

Sidecar just might be the single most versatile feature of ArcGIS StoryMaps. It pairs a stationary media panel with a scrolling narrative panel in a single slide. Story authors can string multiple slides together to create captivating viewing experiences, whether it's a suite of photos with accompanying commentary, a map that readers are guided through one slide at a time, or much more.

Sidecar includes three layout choices, each of which provides useful and engaging narrative choices. The **floating panel** layout is ideal for visually striking media with short captions or descriptions. The **docked panel** layout is optimal for longer narrative content because it doesn't overlap the media. The **slideshow** layout is similar to the floating panel, but readers navigate its slides laterally, rather than vertically, by manually clicking through them. In addition, you can adjust the width and placement of the panel. Panel backgrounds can be black, white, or transparent.

Docked

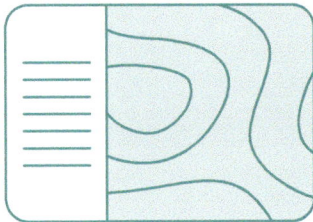

Floating

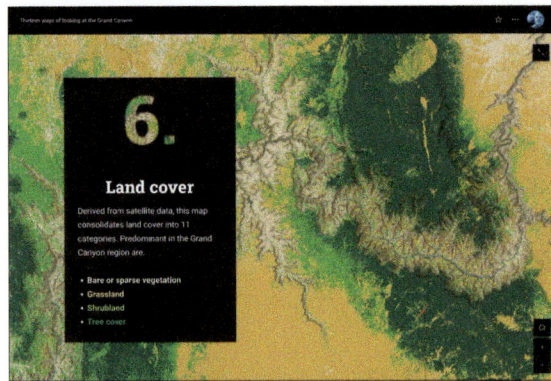

Slideshow

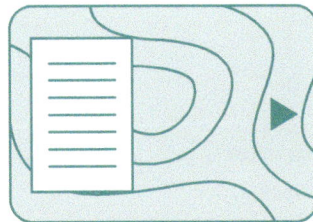

Three layout options for the sidecar immersive block.

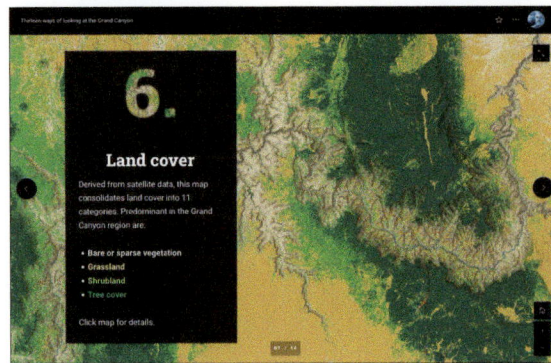

The panels, whether they're floating or docked, accommodate text, as well as photos, videos, maps, and embeds. Media panels can accommodate maps, images, videos, embedded web content, and swipe blocks. The *Getting Started with ArcGIS StoryMaps* tutorial shows you how to create these immersive panels.

A tip: When you add images to the media panel, they automatically fill the space within it (*left*), which means that they're likely to be cropped a bit. You can opt to have the media "fit" the space, which will display the entire media (*right*) but possibly leave some blank space around it.

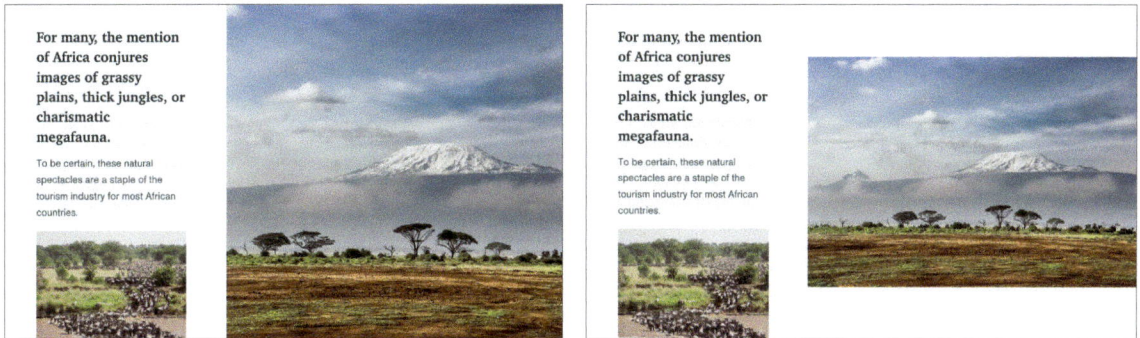

African savanna larger but cropped on "fill" (*left*) versus smaller and full frame on "fit" (*right*).

Sidecars can be particularly useful to intimately intermix maps in your narrative by making them "dance" from slide to slide in what we call **map choreography**. Team member Warren Davison used this technique to good effect in a story about the spotted lanternfly, an invasive insect (see chapter 7). Warren compiled a web map (*next page*) with multiple layers depicting aspects of the invasion. He created a sidecar slide and placed the web map in the media panel, turning on the layer that showed the lanternfly's natural range in China (1). He introduced and described the insect in a floating panel. Then he duplicated the slide, exposing a layer with an arrow representing the lanternfly's arrival in the US as a stowaway (2). He also changed the scale and extent of the map to show both the US and China, and he cleverly moved the floating panel to the middle of the screen. Then he again duplicated the slide, this time displaying layers that highlighted Pennsylvania and Berks County, with a floating panel on the right representing the lanternfly's arrival (3). Finally, he created one more duplicate slide, this time retaining the map's scale and extent but exposing layers that showed the insect's alarming spread across the northeastern US (4).

Elsewhere in the story is another sidecar, a timeline, a map tour—and an

1
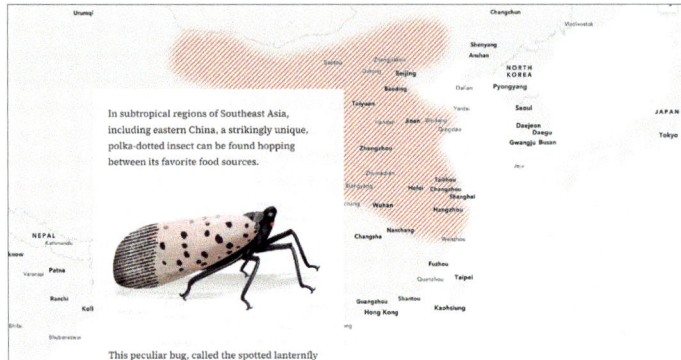

In subtropical regions of Southeast Asia, including eastern China, a strikingly unique, polka-dotted insect can be found hopping between its favorite food sources.

This peculiar bug, called the spotted lanternfly

2
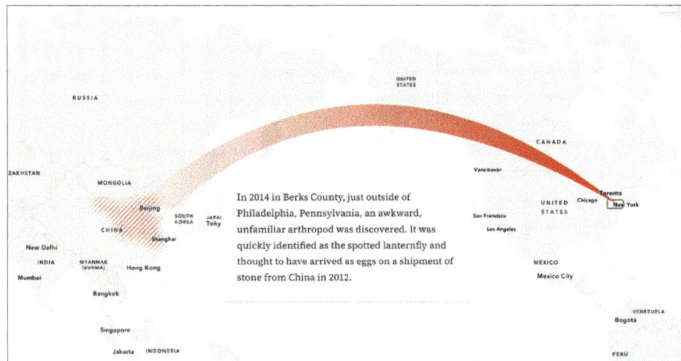

In 2014 in Berks County, just outside of Philadelphia, Pennsylvania, an awkward, unfamiliar arthropod was discovered. It was quickly identified as the spotted lanternfly and thought to have arrived as eggs on a shipment of stone from China in 2012.

3
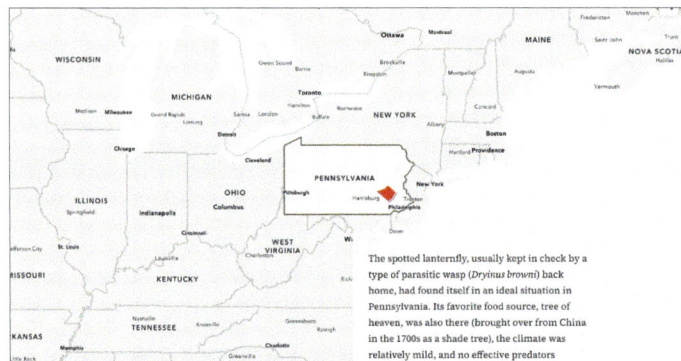

The spotted lanternfly, usually kept in check by a type of parasitic wasp (*Dryinus browni*) back home, had found itself in an ideal situation in Pennsylvania. Its favorite food source, tree of heaven, was also there (brought over from China in the 1700s as a shade tree), the climate was relatively mild, and no effective predators

4
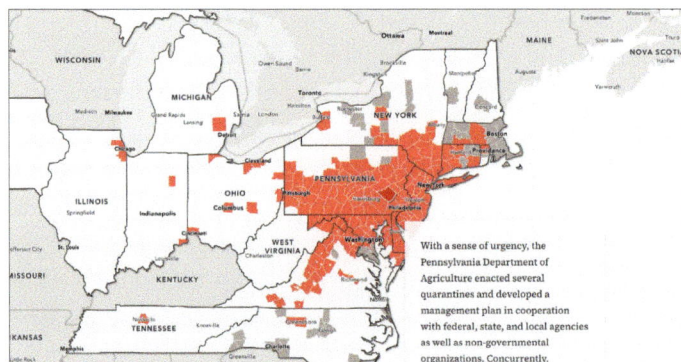

With a sense of urgency, the Pennsylvania Department of Agriculture enacted several quarantines and developed a management plan in cooperation with federal, state, and local agencies as well as non-governmental organizations. Concurrently,

Map choreography: spread of the spotted lanternfly.

animated lanternfly that crawls across the title page. The narrative works as a seamless whole, thanks to team member Heidi Daulton's prose and the use of a consistent color palette and graphic treatments.

Map choreography enables storytellers to seamlessly integrate maps in multimedia narratives and to present and interpret multiple layers of information. Another team member, cartographer Cooper Thomas, has described the various uses of map choreography:

- **Comparing thematic layers within a geography** by keeping map scale and extent consistent but exposing layers serially
- **Comparing data across geographies** by retaining visible thematic layers while changing the map's focal points
- **Recounting an event or sequence** by creating layers representing a sequence—say, a storm track or wildfire extent—and sequentially exposing them in a sidecar section
- **Deconstructing complex maps** by starting with a single layer and revealing additional layers cumulatively, interpreting each new layer as you go
- **Introducing an exploratory map** by interpreting a map in several steps, then presenting an interactive version of the map for readers to explore on their own

An additional advantage of map choreography is the ability to combine a map sequence in the media panel while presenting supporting photos, videos, and especially infographics in the narrative panel.

The sidecar slideshow layout, with its lateral navigation using previous and next buttons, can help relieve scroll fatigue—circumstances in which readers grow weary of long and continual scrolling navigation. I don't regard scroll fatigue as a major issue, except perhaps in very long stories. There's also a risk that your readers will opt not to click through your slideshows or will fail to notice the buttons and scroll by them. You can use this to your advantage by incorporating elements within the slideshow that are useful but not critical to your story. For instance, readers of our **Diversity of Life** ⇦⊃ story can scroll straight through a series of maps of Earth's biomes or move laterally through slideshows displaying representative photos of each biome (*next page*).

All three sidecar layouts offer additional flexibility. In the docked layout, you can place the panel to the left or the right of the media. You can also adjust the width of the narrative panel (narrow, medium, wide). The wide setting splits the browser window exactly in half, which I thought could make for an interesting storytelling format. So I did an experimental story,

**André versus Capability** ↩ (*next page*), about two influential garden designers of the 17th and 18th centuries, André Le Nôtre and Lancelot "Capability" Brown. I placed André's narrative on the left and Capability's on the right. I thought it worked well, but the obscure topic attracted a tiny audience, and I'm not sure anyone noticed my device.

In the floating panel layout, you have the choice of positioning the narrative panel to the left, center, or right. In the slideshow layout, you have three vertical placement options in addition to left, center, or right. In the latter two layouts, you can also make the entire panel transparent, with either light or dark text. That device can work well if the background media isn't so busy that it affects readability.

An additional feature of sidecar allows you to enrich your stories with optional media behavior and map interactions. **Media actions** are toggle buttons or text links that, when clicked, alter the appearance of the media portion of the slide. Media actions can be set up to change aspects of a map or scene, such as layer visibility or extent, or swap out the media entirely for a different item. Media actions are treated as secondary content in a story—in other words, readers aren't required to interact with them. Because of this, media actions are best suited for views of a map or pieces of media content that aren't essential to the narrative but that still add valuable context or information.

Among the many uses of media actions are highlighting an area of interest by having a map zoom to a specific locality, toggling a map layer on or off to illustrate a feature or phenomenon, and showing change over time by turning on temporally specific layers in a web map.

The sidecar's slideshow layout.

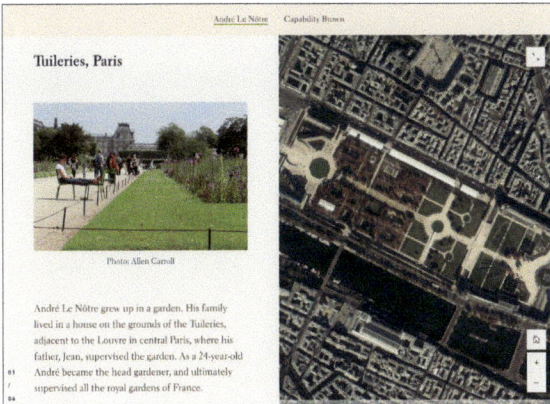

**The Age of Megacities** ⟶ (*below*) uses this technique to demonstrate how the urban footprints of the world's largest metropolises have expanded over the course of a century. For each megacity, a group of media action buttons reveals the city's extent at several points in time.

Media actions can also provide a new perspective on a topic by changing the content of the media panel entirely. No matter what is initially occupying the media portion of a slide, you can configure a media action button that swaps that content for a different map, image, video, or embedded content.

A final note on sidecars: They work beautifully with 3D web scenes. **Answering the Call** ⟶, a story created by my team, includes a web-scene-driven sidecar showing a rescue operation in New Hampshire's White Mountains (*next page*). The scene depicts the rugged landscape and dramatizes the challenges faced by the rescuers. For both maps and scenes,

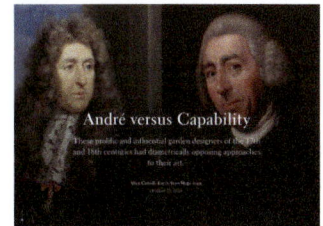

Left and right docked panels in sidecar immersives in *André versus Capability*.

*The Age of Megacities* uses media actions to show urban growth over time.

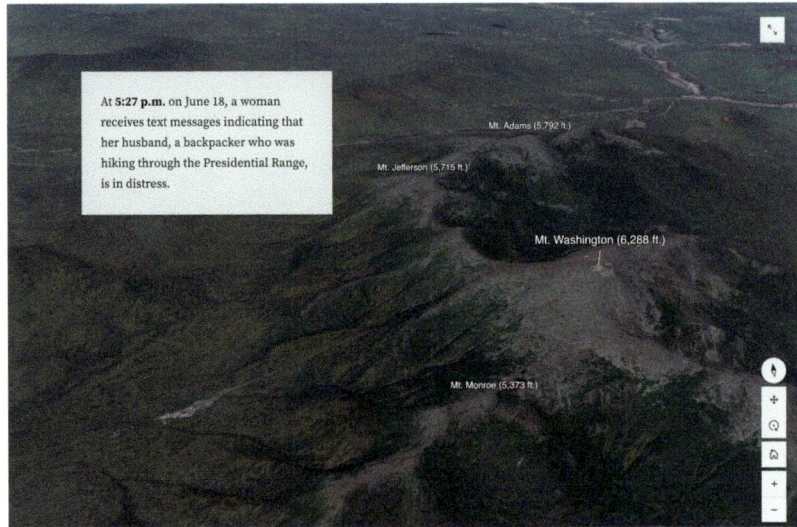

At **5:27 p.m.** on June 18, a woman receives text messages indicating that her husband, a backpacker who was hiking through the Presidential Range, is in distress.

Mt. Adams (5,792 ft.)

Mt. Jefferson (5,715 ft.)

Mt. Washington (6,288 ft.)

Mt. Monroe (5,373 ft.)

Sidecar featuring a web scene.

it's best to use a single web map or scene within a sidecar series, which ensures that, as readers scroll, they'll see a smooth transition rather than a momentary pause or jump as a separate map or scene loads.

## Map tour

The other ArcGIS StoryMaps immersive element, map tour, allows you to plot points on a map and add media content and narrative text that will display in a side or a floating panel, depending on your layout choice. There are two types of map tour: **guided tour** and **explorer tour**. Each of these has two layout options of its own.

**Guided tour** lets you walk your audience through a set of places in sequence. As readers scroll, the map's focus shifts from one tour stop to the next; accompanying text and media (photos and/or video) for that place are displayed alongside it.

Two layouts are available for guided tours. The **map-focused layout** puts images and descriptions in floating panels, with the map occupying the remainder of the screen. The **media-focused layout** puts map and text in a side panel, letting your images become the stars of the show. In both layouts, the map is interactive, so readers can explore on their own. Many authors choose the map-focused layout; I usually prefer media-focused, which devotes more real estate to the images and lets the map play its more modest narrator role without dominating the screen (see chapter 6).

**Explorer tour** offers a less linear experience. Points are plotted on the map, with a corresponding gallery of thumbnails in either list or grid form

Guided tour: map-focused

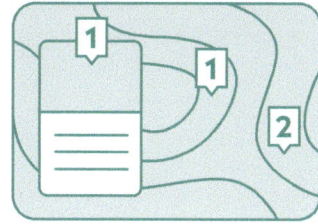

Guided tour: media-focused

Explorer tour: list view

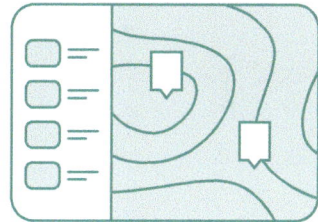

Explorer tour: expanded view

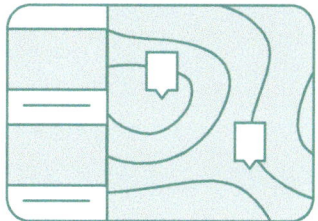

Guided and explorer tour options.

Guided tours using the sidecar immersive style. *From top*: floating, slideshow, docked.

on the side panel. Clicking any item on the map or the side panel reveals its associated media or narrative information. The upshot is that readers can browse the points in whatever order they prefer.

A tip for tour authors: You can—and in most cases should—manually set map zoom levels for each point, or stop, in your tour. We frequently encounter tours in which map scales are too large or too small to properly orient readers. There's no single, ideal scale; often it's useful to set the initial tour point's map at a smaller scale so readers are oriented to the regional setting of your tour and then zoom in for subsequent points.

One of the things I enjoy most about ArcGIS Story-Maps is the flexibility it offers. The images on these pages are all made from an instructional story, **Six Scottish Hikes, Six Tour Formats** ⊝, that I authored about a family vacation in Scotland. Several of the formats used the map tour immersive format (*previous page*); additional formats use sidecar (*left*). The sidecar versions required some additional work, namely creating a web map and putting each tour stop and route segment into a separate layer of the map. I then sequentially exposed the points and segments in the narrative. For the sidecar stories, I used the floating panel, docked panel, and slideshow layouts to create tours documenting three of our hikes on Scotland's Isle of Arran.

We strive to keep the story-authoring process as intuitive as possible while providing a broad array of capabilities. The balance of simplicity and feature richness is challenging to maintain, but the result (we hope) is an array of tools and solutions that enable you, as authors, to come up with creative solutions to storytelling challenges.

The sidecar and map tour immersives provide flexibility in determining the overall structure of your story. Some stories may consist almost entirely of a sidecar or map tour—except perhaps for a cover page and a concluding paragraph or call to action. Others may insert several map tours and sidecars within an overall scrolling experience. Still others may include no immersive sections at all. If you're planning a virtual walking tour of a historic district, a map tour may be all you need. My team's series

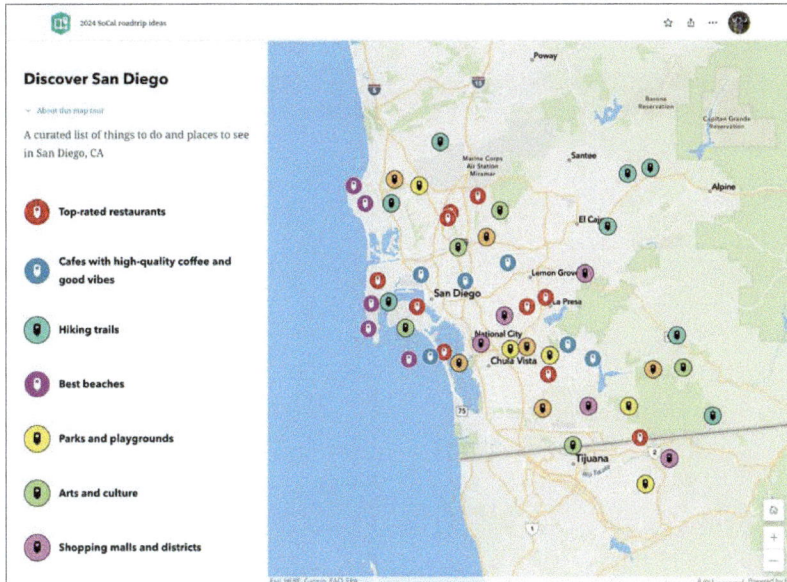

Categorized map tour.

of stories on the Anthropocene, *Living in the Age of Humans,* uses a rich mix of sidecars, tours, swipes, timelines, and other devices. It's useful to keep these options in mind as you plan your stories.

A relatively new feature of map tour is the option to separate points of interest into categories that can then be viewed separately. This provides a function similar to the Shortlist app, which was released early in my team's tenure but was discontinued (see chapter 5).

## Swipe

Swipe enables your readers to move a slider left and right to compare two related maps or graphics. It's an elegant way to examine contrasting views of the same geography, including related thematic maps or change over time, as in the swipe on the next page juxtaposing an 1851 map of the heart of Washington, DC, with a contemporary imagery view.

Swipes engage your readers—or at least some of them. Not everyone will want to manually manipulate elements of your story. An alternative—which I prefer in most circumstances—is to use the sidecar immersive style to choreograph a pair or series of images or maps to appear in sequence. No need for readers to grab the left–right handle to make the comparison; they can scroll up and down in the sidecar to compare.

Back in the classic Story Maps era, Esri released a swipe function as a stand-alone app. We also released an app that combined a swipe and a spyglass function that featured a small, moveable window that allowed

Swiping to compare maps and images of Washington, DC.

The old Spyglass app.

readers to peer through one map into another. It was less useful than the swipe, but it had a peculiar, voyeuristic appeal that people seemed to love. The Smithsonian's website featured then-and-now spyglass maps for a half-dozen US cities, and the stories went modestly viral.

## Timeline

ArcGIS StoryMaps has a timeline feature that supports temporal sequences. It has two formats: **Waterfall** centers the event information and stacks items on top of one another, and **single-side** aligns most information to the right of a vertical line. Timeline type and colors conform to the design theme chosen by the author.

The waterfall timeline (*next page, left*), **Bombing Missions of the Vietnam War** ⊖, describes major bombing missions of the Vietnam War. The timeline function (*next page, right*) needn't be used solely for temporal matters. The image shows a story by Aaron Koelker called **Green Oranges & Land** ⊖, with a **single-side** list of HLB mitigation measures. HLB is an abbreviation for huanglongbing, or citrus greening, a bacterial disease spread by an invasive Asian insect.

These timelines hint at the enormous range of topics that storytellers feature among the hundreds of thousands of narratives produced every year by the ArcGIS StoryMaps community.

**Operation Barrel Roll (1964–1973)**

Over the course of the Vietnam War, more than 2 million tons of bombs fell on Laos alone. Most of those munitions targeted the Plain of Jars, an archaeologically significant landscape harboring communist Pathet Lao insurgents. Per capita, Laos remains the most heavily bombed country on earth. But nearly a third of these bombs failed to detonate on impact, and nexploded ordnance continues to maim and kill scores of Laotian civilians each year.

**Operation Rolling Thunder (1965-1968)**

Operation Rolling Thunder, the United States' sustained bombing campaign against North Vietnam during the mid-1960s, aimed to demoralize and degrade the country's government and people. The campaign was blunted by rigid targeting restrictions—in some cases, imposed directly by President Lyndon B. Johnson. It was ultimately a strategic failure, and only exacerbated antagonism toward the U.S.

**Operation Steel Tiger (1965-1968)**

Operation Steel Tiger targeted the Ho Chi Minh Trail, which served as the North's primary supply line into South Vietnam. Snaking through the rugged terrain of neighboring Laos and Cambodia, the trail allowed North Vietnamese forces to covertly move personnel and supplies with relative impunity. Despite persistent attempts to destroy the supply line through saturation bombing, the trail operated almost continuously until the end of the war.

**Operation Menu (1969-1970)**

In spring 1969, the United States began a secretive bombing campaign in eastern Cambodia, conducted without the knowledge of the American public or Congress. After a year, government whistleblowers exposed the

**Protective Covers**

Fine mesh netting can prevent the psyllids from accessing young trees, ensuring a strong, healthy start through the period when the trees are most vulnerable and increasing their chances of survival into the fruit-bearing stage[6].

**Reflective Mulch**

Applying reflective mulches around the base of the tree (akin to large sheets of aluminum foil laid flat on the ground) disorients flying psyllids by reflecting UV light from below and can deter them from settling on trees[7].

**Pesticides**

Although they come with a gamut of environmental concerns, chemical pesticides remain an effective way of limiting HBL infection by controlling populations of psyllids within groves.

**ENPs and Microbe Control**

Focusing on soils, Enhanced Nutritional Programs (ENPs) use timed-release fertilizers to help trees mitigate deficiencies caused by HLB[8]. A second area of research seeks to increase nutrients by promoting the natural interactions of microbiomes found around the trees' roots[9].

**Trunk Injections**

Injecting antibiotics like oxytetracycline hydrochloride directly into the tree trunk can help citrus withstand the effects of HLB. This is done with a syringe-like device and leverages the tree's vascular system to disperse the compound

Timeline layout options.

## Adding audio

Sound is generally underappreciated as a storytelling device. Audio can add an extra dimension to your stories as background or ambient sound and as click-to-play elements. For the latter, choose **Audio** in the block palette and upload an MP3 or WAV file. You can also embed directly from supported third-party audio hosts.

Two of my favorite uses of sound in multimedia stories: providing atmosphere with ambient sound and giving voice to an individual. The Charles River Watershed Association paired contrasting photos and audio of a stagnant, quiet impoundment above a dam site with the roar of rushing water (an erosion hazard) downstream from a dam. Web browsers require readers to make an initial choice to unmute background audio within a story, after which all background audio will play automatically as the reader scrolls.

The Grand Canyon Trust profiled several tribal people describing their deep spiritual ties to the canyon. Portrait photos, text, and click-to-play recordings painted a vivid picture of the canyon as both a natural attraction and a rich cultural landscape.

## Embedding

One of the most versatile features of ArcGIS StoryMaps is its **embed** function. Almost anything on the web can be embedded in stories: tweets, audio tracks, music playlists, interactive charts, ArcGIS apps. You can embed photos if they're hosted elsewhere—for example, on Flickr or your own server. But we strongly recommend uploading your photos directly to the story. ArcGIS StoryMaps does some behind-the-scenes magic to optimize your images to help them load quickly. You can adjust your photos using the builder's editing tools. You can also upload videos with a maximum file size of 50 MB. Beyond that, embedding from YouTube or other services is the way to go.

Embeds can also integrate functions provided by other ArcGIS apps. For instance, ArcGIS Survey123 forms can be embedded in stories to create simple crowdsourcing instances (*below*). My team produced a **Share Your Earth-Places** ⬡ story in April 2024 to celebrate Earth Day. We invited people to

Sharing earthplaces with an embedded survey.

share their favorite places in nature by filling out an embedded form with a photo, location, and description. Contributors click Submit and refresh the page, and the new submission appears on a 3D globe alongside hundreds of others.

Embeds can be used to incorporate interactive charts in your stories. Team member Cooper Thomas has used this technique in several stories, including *Bombing Missions of the Vietnam War*. He embedded the chart (*right*) he created using ArcGIS Dashboards as a single bar graph. Cooper has also used open-source graph libraries but likes the Dashboards app because it can tap the same data sources as the maps in his story.

Embeds present many opportunities for enriching your story, but you'll want to consider multiple variables, including the types of embedded media, how embeds are displayed, and where in your story you add them. Refer to the *Getting Started* story and links for more information.

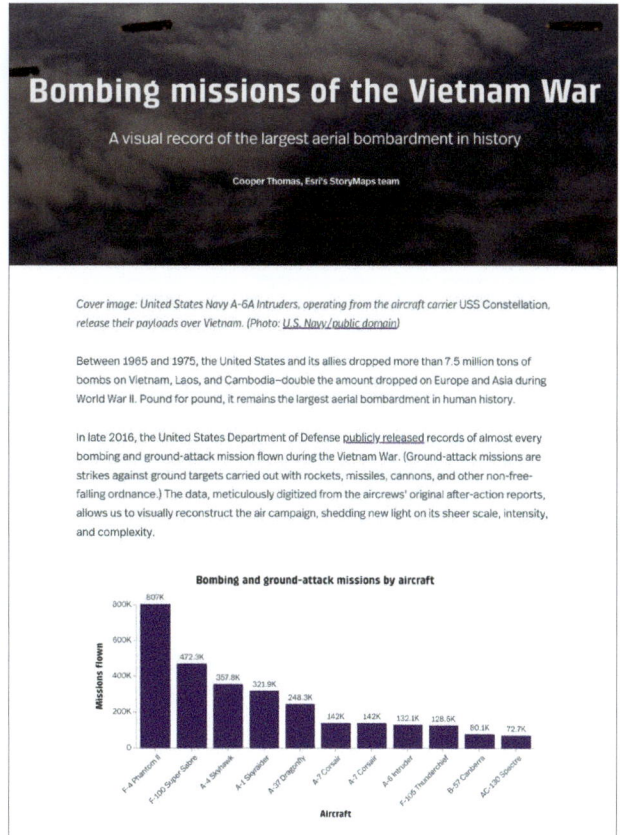

Embedding ArcGIS Dashboards graphs.

## Refining your design

The ArcGIS StoryMaps builder gives storytellers an almost infinite array of design options, including cover formats, navigation controls, header treatments, logo display, color palettes, and typography. To minimize confusion, controls for design elements are presented and organized into separate functions.

The first item in the **design controls**, located on the header bar in the builder, provides many **cover** formats and options (*next page*). The default cover format is minimal (*top left*), with the title appearing below an optional image. You can experiment with side-by-side and full-screen options; whatever you choose can be changed at any time. I usually find myself gravitating toward the full-screen option because of its greater visual impact (*top right*). But finding an image that works with the title panel can be challenging. Between the cover layout options and the panel appearance controls, you have plenty of choices, including position and

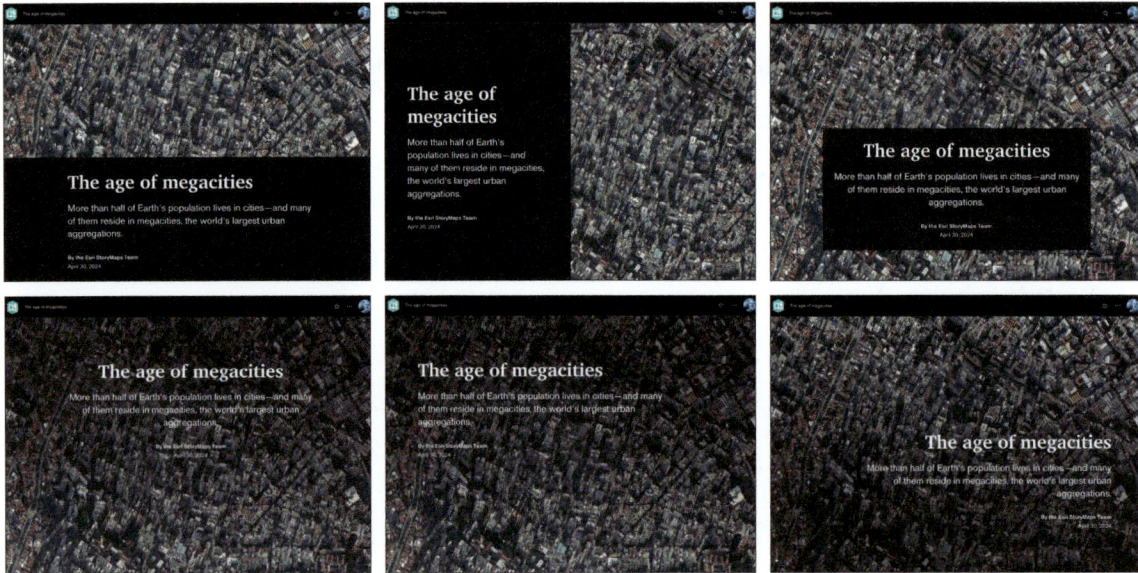

Cover options.

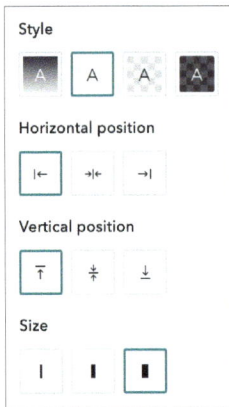

Panel appearance options.

alignment of cover text and whether it appears within an opaque panel or floats atop the image (*bottom row*).

You can use **optional story sections** to create a navigation bar in your story by turning on **navigation**. Any header in your story will appear in a bar that's visible throughout the story. You can edit the text in the navigation bar without changing the headers in the narrative itself. A **credits** section is on by default. Credits appear at the end of the story—a great place to cite sources and provide links to data resources. You can switch off credits if you don't need to credit your sources or if you're doing so individually for maps and images within the story.

The design panel's anchor tenant might be the **theme** functions, which let you transform the look and feel of your story with a single click. Themes provide customized combinations of colors and typography with designs for buttons and other story elements.

Themes can be accessed three ways. First, buttons for six predesigned themes are directly accessible in the Design panel. We've created these themes to support a variety of narrative and editorial approaches. Second, the Featured Themes Gallery button provides additional options.

Finally, you can create your own custom themes by clicking the Themes button on the ArcGIS StoryMaps home page. A separate builder function allows you to choose among almost unlimited color palettes for background, typography, buttons, and quotes. Type choices are legion and include Google's selection of hundreds of fonts. You can choose from

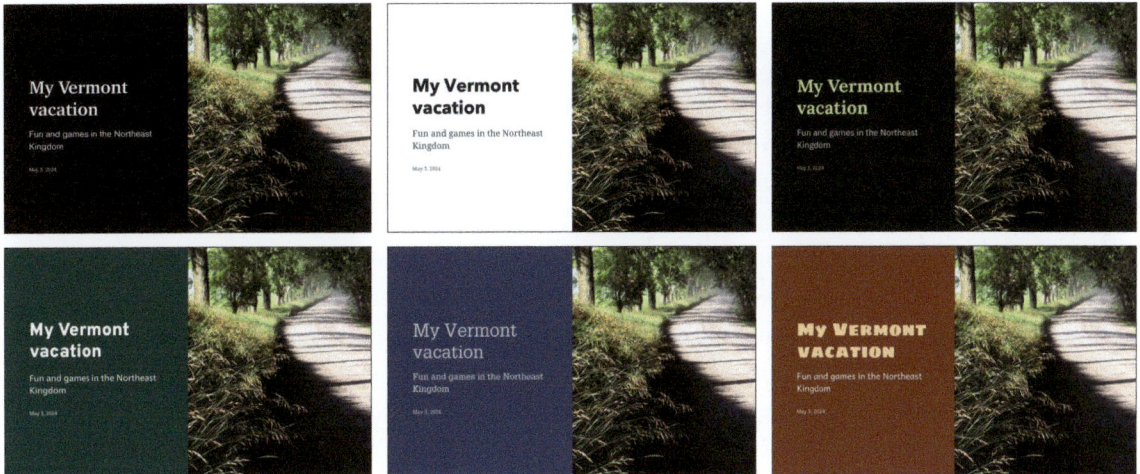

A sampling of story themes.

several design options for quotes, links, buttons, and separators. Choosing a custom theme will also alter the appearance of your express maps.

Why create a custom theme? It's a great way to give your story a distinctive look and feel. It's also a way to help set a mood for your narrative. More somber or nocturnal stories might lend themselves to dark backgrounds. Stories about geology or nature might beg for earth tones or greenish hues. Typefaces can look more or less formal or child-friendly. Using serif or sans serif typefaces may lend your story a historical or contemporary tone.

It's best to be somewhat restrained with type and color choices; gaudy palettes and cutesy fonts risk distracting your readers and reducing legibility.

Custom themes can be created to align with organizational branding and design guidelines. ArcGIS account administrators can make brand-consistent themes available across their organizations so that stories published by multiple teams and authors all have a consistent look and feel.

Need help making type choices? Search the web for "Google font pairings" and you'll find several sites with tasteful and attractive suggestions. I confess to having been worried that providing near-infinite design options to our storytelling community would lead to countless stories with garish colors and tasteless typography. I needn't have been concerned. Most authors have shown admirable restraint. It's exciting to encounter myriad stories with restrained choices that align with organizational brands and create immersive and visually appropriate user experiences.

Another, more direct way to brand your stories is to include your organization's logo. The design panel allows you to add a logo to the header bar or to your story's cover.

## Step 3: Polish

My team and I often find that the final 5 percent of work to polish our ArcGIS StoryMaps stories makes a dramatic difference in the quality of the stories we produce. Sometimes a late and relatively simple change to the styling of a story or the addition of, say, a custom separator will add visual charm to a narrative.

An example: I had nearly finalized **Thirteen Ways of Looking at the Grand Canyon** ⊝, featuring a selection of thematic maps. Team member Warren Davison suggested some visual fun with the numbers that appear in the story. He created numbers as artwork, sampling each map layer to fill the interior of the numerals. It's easy to assume that tweaks like this are merely decorative, but, as I've stressed before, creating a distinctive and consistent look and feel adds visual panache to your stories and makes the reading and viewing experience more pleasurable and perhaps more memorable. This applies to all elements of your story: color palettes, font choices, map and infographics styling, buttons, separators—even quotations and hyper-links, for which the theme builder gives you several options.

It's standard practice for us to ask one or more members of our team who have not been involved in a story project to give the nearly complete

*Thirteen Ways of Looking at the Grand Canyon: adding a unifying graphic element.*

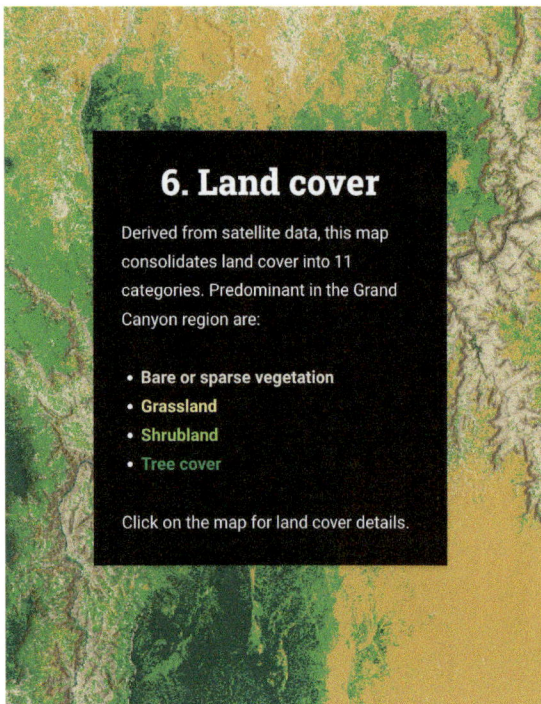

### 6. Land cover

Derived from satellite data, this map consolidates land cover into 11 categories. Predominant in the Grand Canyon region are:

- **Bare or sparse vegetation**
- **Grassland**
- **Shrubland**
- **Tree cover**

Click on the map for land cover details.

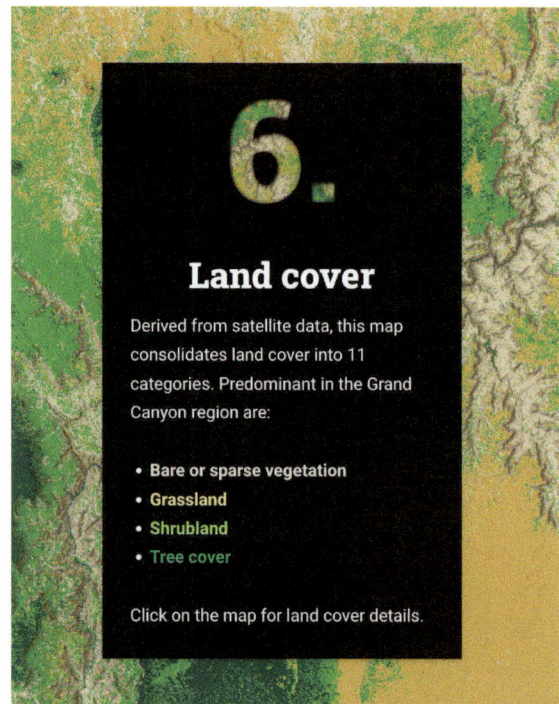

### 6.
### Land cover

Derived from satellite data, this map consolidates land cover into 11 categories. Predominant in the Grand Canyon region are:

- **Bare or sparse vegetation**
- **Grassland**
- **Shrubland**
- **Tree cover**

Click on the map for land cover details.

narrative a careful review. We post our stories on a digital whiteboard in our shared workspace and welcome comments and ideas from team members. We have someone on the team copyedit the story to correct typos and catch inconsistencies. As author John McPhee has said, typos are as hard to catch as cougars. He's right. They can slip by the sharpest eye. They're less dangerous than cougars, fortunately, and easier to fix after the fact in web-based media, where a story can be republished instantly.

But late-stage reviews of text shouldn't be limited to correcting typos. It's useful to take a fresh look at the style and tone of the writing. Is the text appropriate to the audience? Does it lapse into jargon or use technical terms that some readers may not understand? Is the text clear and concise? Are any paragraphs too long? Long blocks of uninterrupted text look forbidding and can scare casual readers. Does the style of the text align well with the look and feel of the story? Does it fit the theme?

If you're writing about your GIS work or other specialty but targeting a general audience, you might try a technique I've used myself—think of your mom. When you talk to your mom about your work, you don't dumb things down; you respect her intelligence but avoid using technical terms, such as "hosted feature service," "geodatabase," or "cadastral."

A key step in the polishing process is ensuring that your story is **accessible** to readers with visual and other impairments. The primary task in this process is writing alternative text, or alt text. Read aloud by machine screen readers, alt text describes the look of the maps and other visuals and embedded content in a way that captions don't. For example, a caption might supply the name of a map, but alt text will describe the colors or the elements of the map (such as a north arrow or a scale bar) so that sight-impaired readers can get a sense of how the map's meaning in the story is visualized. Our designers and developers have worked hard to make ArcGIS StoryMaps as accessible as possible. In addition to facilitating the easy addition of alt text, the ArcGIS StoryMaps builder alerts you if you're choosing a text color that, for instance, provides inadequate contrast to the background. Note the "Not legible" warning at the bottom of the color picker image (*right*).

Another vital step in the polishing process is ensuring that you have secured all the necessary **rights and permissions** for the elements of your story. Receiving permissions and citing sources may be less vital for a student project published to an organization than a story shared with the world. Obtaining permissions can be a tedious and time-consuming task, so it's best not to leave this process to the end of your production effort.

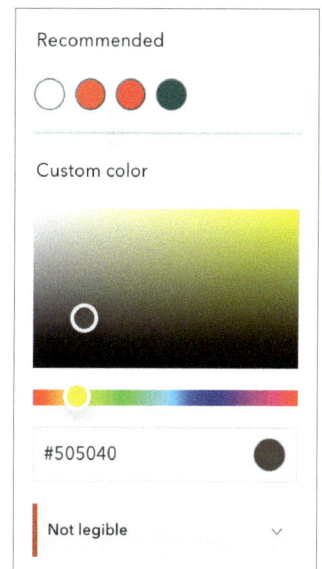

Color picker with a legibility warning.

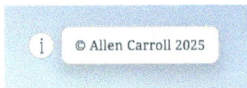

Image credit information appears on hovering.

You can reduce the level of effort for permissions by seeking rights-free images and other content as you build your story. Several rights-free image sources are available on the web, including Wikimedia Commons and Unsplash.

You can add **attribution** and alt text using the toolbar that appears when you hover over images and other items in the builder. Once you add attribution information to a photo, a small "i" icon appears in its upper-left corner; credit text becomes visible when readers hover over the icon.

# Step 4: Publish

Ready to publish? The mechanics of publishing are relatively simple and can be completed in just a few minutes. The *strategy* for publishing is another matter. I'll briefly touch on the mechanics, and then dwell a little more thoroughly on strategy.

Not surprisingly, you'll want to click Publish in the builder's header bar. That will open a **Publish Options** menu (*below*), which has two panels. On the left are options for configuring the **story card**. Here you can determine the thumbnail image, story title, and a brief summary that will show, for instance, when the story is shared on social media or comes up in a search result.

The Publish Options page in the ArcGIS StoryMaps builder.

Story details

Configure what people see when your story appears on ArcGIS, social media sites, and in search engine results

Share

Who can see this after it is published

Set sharing level

Organization

Set group sharing

Search groups

Advanced options

☐ Allow duplication

☑ Show in web search results

**Thirteen ways of looking at the Grand Canyon**

A cartographic tour of a natural wonder

✎ Edit

To review display, analytics, and language options, go to story settings

On the right are the sharing settings. **Private** makes the story visible only to you. **My organization** makes it accessible to other people in your ArcGIS organization. **Everyone** makes your story public. You can also share your story to ArcGIS Online groups you're a part of by using the search box. If you're working for a large organization but creating a story for a wider audience, you may want to publish to your organization first and let your colleagues know that your story is available for review.

Below the sharing settings are two advanced options. **Allow duplication**, which is turned off by default, gives anyone to whom your story is visible the ability to create a copy of the story in their own account. **Show in web search results** is on by default and determines whether your story will appear in Google and other search engines. Disabling this option is useful if you want someone outside your organization to read or review your story before it's easily accessible to the public.

When you're satisfied with these settings, click Publish again in the upper-right corner. As part of the publishing process, a behind-the-scenes story checker will look for permissions issues with the maps in your story and flag any maps or layers whose sharing levels are more restrictive than that of your story. After you click Publish, a screen appears with options to copy the link, share to social media, scan a QR code to view on your smartphone, view the published story, or edit it.

It might be tempting to hope that if you build it—and publish it—they will come. Alas, we are all awash in an ocean of content. Unless you're a celebrity or a veteran influencer, your story, no matter how glorious or groundbreaking, will almost certainly be doomed to obscurity—unless you come up with a plan. If you're fortunate, you have access to an experienced communications specialist with whom you can refine your publishing strategy.

I'm fortunate. Team member Michelle Thomas is a veteran creator of story programs, having produced three pioneering story campaigns: *Travel Tuesdays* and *My Public Lands Summer Road Trip* for the Bureau of Land Management and *Fridays on the Farm*, a weekly profile of farm families using our classic platform for the US Department of Agriculture. She joined our team in 2019 and has continued to polish her considerable skills. The items listed in this section summarize her strategy for rolling out a new story.

Michelle counsels story authors to think from the start about their intended audiences and how those audiences can be reached. Are the story's hoped-for readers part of a community that connects online through an organizational website or social media hashtag? Does the audience

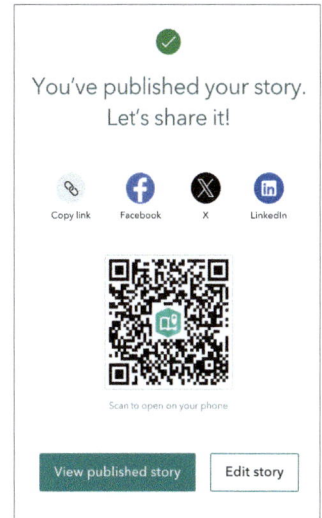

You've published your story. Let's share it!

Copy link    Facebook    X    LinkedIn

Scan to open on your phone

View published story    Edit story

Sharing options: the final step in the publishing process.

gather at a key event around which you can time your story's publication? Are there holidays or anniversaries related to your story that you can use to attract readers?

Whether you're collaborating with colleagues in your organization or outside it, Michelle advises that you summarize your communication strategy in a document or presentation to share with your colleagues. She advises to include the following:

- A brief description of your story
- A list of key messages
- A call to action—what you intend to have your audience do in response to your story
- Blog posts, videos, or other online content related to your narrative
- A list of social media platforms and handles appropriate to your audience and topic
- Links to shareable content that others can use in social media
- A list of dates for events and action items, including when to post to social media
- One or more sample social media posts with ready-made media assets, such as maps, photos, and videos, to make sharing easy for all stakeholders

Basic to Michelle's strategy is what we call "hub and spoke." The idea is to make your story the centerpiece of several related items, each of which can increase its impact and attract readers. For example, we'll prepare a post to run on Esri's popular ArcGIS Blog that describes the story or perhaps explains how the story was produced—and that links to it, of course. Or we'll make a short instructional video that describes how to use one of the functions employed in producing the story.

Finally, Michelle urges story authors to keep an eye on how the story is performing. "Most social media platforms have built-in analytics," writes team member Ashley Du, who works closely with Michelle. "With ArcGIS StoryMaps, you can add your own Google or Adobe Analytics codes into your stories. This approach is particularly helpful if you have Google or Adobe Analytics for your website, social media campaigns, and other digital platforms."

# Step 5: Preserve

I organized this chapter around four steps: Plan, Produce, Polish, Publish. But there is a fifth step, which involves maintaining the health and relevance of your story. It's worth expending a little ongoing effort to *preserve* your story.

First, every few weeks or months, you should pay your story a visit to make sure all the maps and other multimedia elements are alive and well. Open your story in an incognito browser window; this will show you whether readers are being asked to sign in if a map or layer isn't publicly accessible.

Check your story's analytics: If it's getting a lot of visitors, congratulations. If not, it may be worth a second round of social media posts or a new strategy to attract new readers. The messaging for your original push may not be resonating with your audience. Try new words and key phrases. Use new hashtags.

If your story gets a steady stream of visitors, it may be worth updating the story with new data. On occasion, I've been guilty of publishing a story and then getting excited about other projects and neglecting my older opuses. Think of your stories as living things that need a little care and feeding.

Finally, keep your eyes and ears open for opportunities to promote your story a second time. An item in the news may be related to your story's topic; a recurring event or anniversary may provide a good excuse to repost to your social media accounts.

If you're new to ArcGIS StoryMaps, I hope this chapter has whetted your appetite to start creating your own stories. If you're a veteran, perhaps you've gained a few insights or learned some new approaches. It's fine to read about creating stories—and I'm grateful that you've taken the time to peruse my book—but the best way to develop your skills is to plunge in and start experimenting. Over the years, I've found that creating outstanding stories is hard work. It's messy, iterative, best done collaboratively, and—as with all skilled labor—something you'll get better at with practice. It's also rewarding to see your ideas take shape as informative and inspiring multimedia stories. Welcome to the ranks of storytellers who are using this medium to open people's eyes and minds to the world—or to various corners of it, no matter how local or obscure. Join me in sharing and experiencing the joy of telling stories with maps.

# The Ebony Project

From understanding a species to saving it through community-based tree planting

**Author:** The Ebony Project

**Medium:** ArcGIS StoryMaps collection

**Story behind the story:** Bob Taylor, cofounder of Taylor Guitars, has a passion for wood—especially ebony, native to equatorial Africa and valued as ideal for the fretboards of acoustic guitars. In 2016, following his purchase of an ebony mill in Cameroon, Taylor launched the Ebony Project to learn about the biology, ecology, and human impacts of the tree.

**Why it's special:** Bob formed an alliance with Tom Smith, a professor at UCLA who had helped establish the Congo Basin Institute. The partnership resulted in new knowledge about ebony trees. Renovation of the mill and tree-planting programs benefited local communities and the Congo Basin rain forest— second largest in the world—as well.

An ArcGIS StoryMaps collection featuring the project opens with a video of Taylor, who provides background and context. Three narratives describe the history, origin, and growth of the project, and a final section summarizes its benefits and alignment with global sustainability goals.

esri
THE SCIENCE OF WHERE

Collection

## The Ebony Project

Tag along for this journey as we go through the collaborative efforts of Taylor Guitars and the Congo Basin Institute.

SEPTEMBER 2024

In 2016, Taylor Guitars launched The Ebony Project to conduct basic ecological research on a traditionally used tonewood and to plant ebony and fruit trees in several small villages that buffer a U.N. World Heritage Site in Cameroon.

Implemented by The Congo Basin Institute, the Project has produced landmark independent research and developed a vibrant community-based planting and agroforestry program. As of September 2024 the project has planted 40,000 ebony trees and 20,000 fruit trees. Planting continues.

This StoryMap is an Annex to the Ebony Project's 2023 Annual Progress Report, which can be found, along with other project documents, at crelicam.com/resources.

An Introduction from Bob Taylor

I. History

II. Genesis

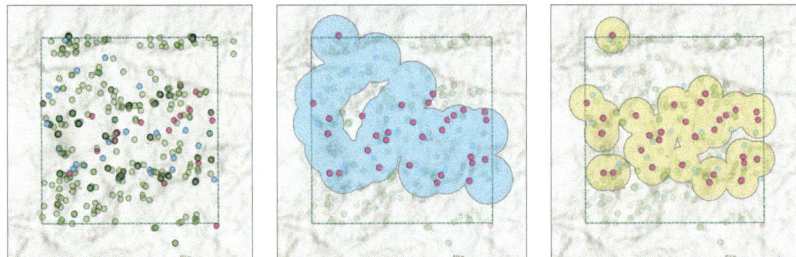

*Top*: Video and stories become chapters in a larger narrative, presented in an ArcGIS StoryMaps collection.

*Above*: A study area census maps male and female trees, pollenation ranges, and small mammal seed dispersal.

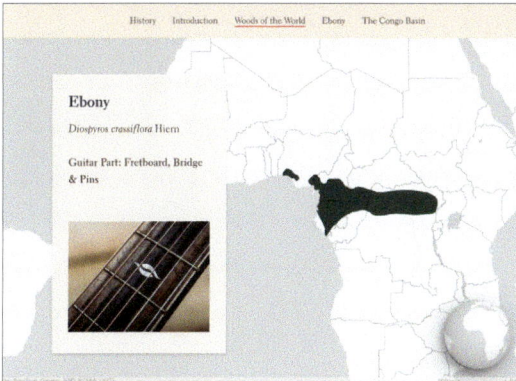

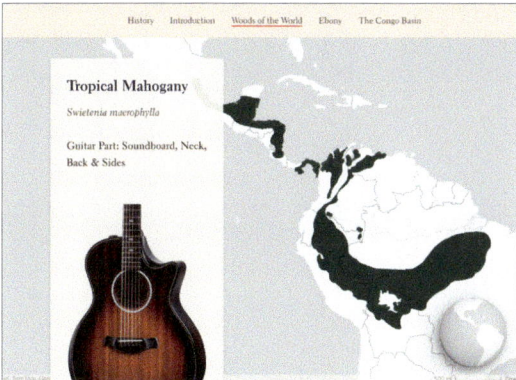

*Top*: The Ebony Project team on a scientific mission in Cameroon.

*Left*: Two of several range maps of tree species used in guitar manufacure.

*Above*: A community member tends an ebony sapling in a local nursery.

# Global Fishing Watch

## Illuminating vessel activity on the open ocean

**Organizations:** Global Fishing Watch and Esri's ArcGIS Living Atlas team

**Medium:** ArcGIS StoryMaps

**Story behind the story:** Global Fishing Watch (GFW) is a nonprofit organization that makes data on human activity at sea publicly accessible by providing a global, real-time view of fishing. The group's fishing intensity layers are available in ArcGIS Living Atlas of the World.

GFW uses innovative detection tools to classify vessel activity. The routes themselves are tracked through positioning systems that report participating ship locations several times a minute. These trails can then be analyzed and classified by apparent activity.

ArcGIS Living Atlas team members Craig McCabe, Emily Meriam, and Keith VanGraafeiland created a story featuring a series of maps that show how this dataset can be visualized and combined with other data to reveal patterns and trends.

Intended for a professional audience, the story combines technical depth with stunning cartography.

**Why it's special:** World fish consumption is increasing roughly 3 percent annually—almost double the population growth rate of 1.6 percent. GFW's data is key to understanding fishing patterns and seeking sustainable fisheries in the future.

*Above:* Industrial fish trawlers operate in the seas off Myanmar.

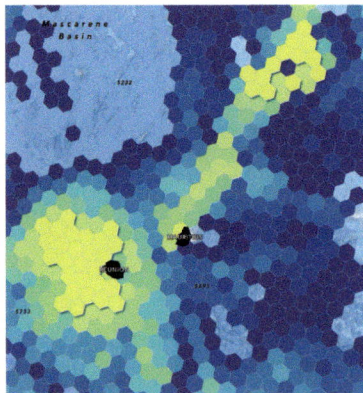

Equal-area hex bins show the number of months per year fishing activity is occurring. Yellow is year-round, darkest blue equals one month.

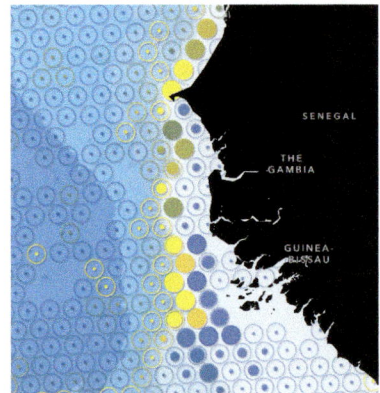

Here, fishing intensity is compared with the slope of the underlying seafloor. The color of the symbol represents steepness of slope, while the size of the interior circle indicates level of fishing intensity. Fish tend to gather at the sloping edges of continental shelves.

Much of the fishing activity in Europe is taking place in the shallow waters around Norway, Denmark, the United Kingdom and Ireland, France, Spain, and Italy.

The Nystrøm, pausing through Brønnøysund harbor, fishes using purse seines and is registered in Norway.

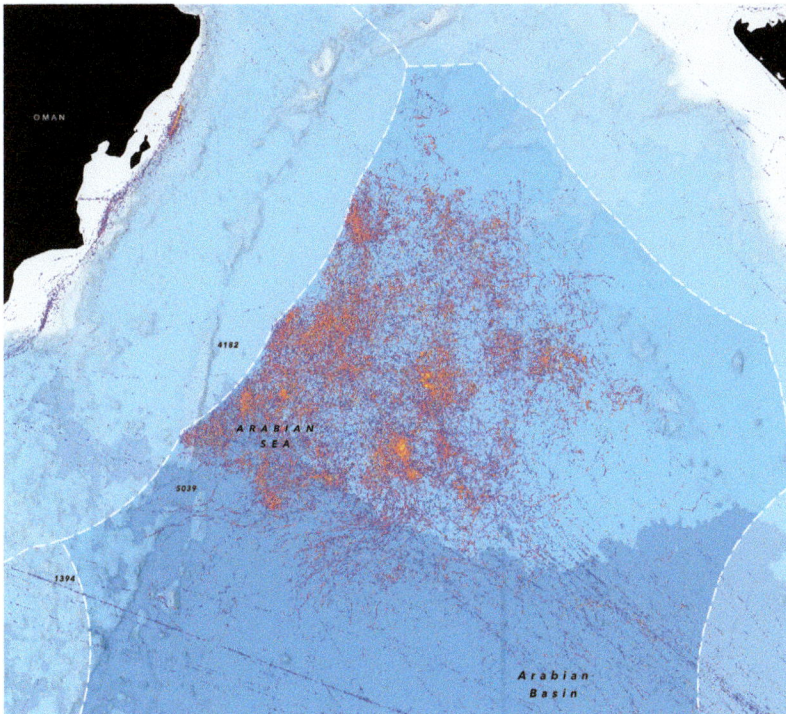

*Above*: Intensive fishing in the coastal waters of western Europe.

*Left*: Patterns of fishing can be dramatically influenced by invisible ocean boundaries. Exclusive Economic Zones (dashed white lines), including Oman's, constrain fishing activity in the Indian Ocean.

# Epilogue

## To inform and enable

As I get close to finishing the happy task of writing this book—and as I approach the end of a long career—my primary emotion is gratitude. It has been a great privilege to be associated with the two organizations which, in my view, best epitomize the power of maps and geography to increase our understanding of our dynamic world.

I worked for the National Geographic Society for most of three decades. During that time, I was honored to help carry forward and modestly enhance a century-old tradition of superb cartography. I played a small role in creating dozens of large-format wall maps that hang in countless classrooms and home offices, helping produce two editions of the society's *Atlas of the World* and nudging the society toward using maps in new ways as the digital age dawned.

At Esri, where I have worked for 14 years, I helped create web-based, place-enabled storytelling capabilities that have proven to be of value to

Terrestrial ecosystems of the world.

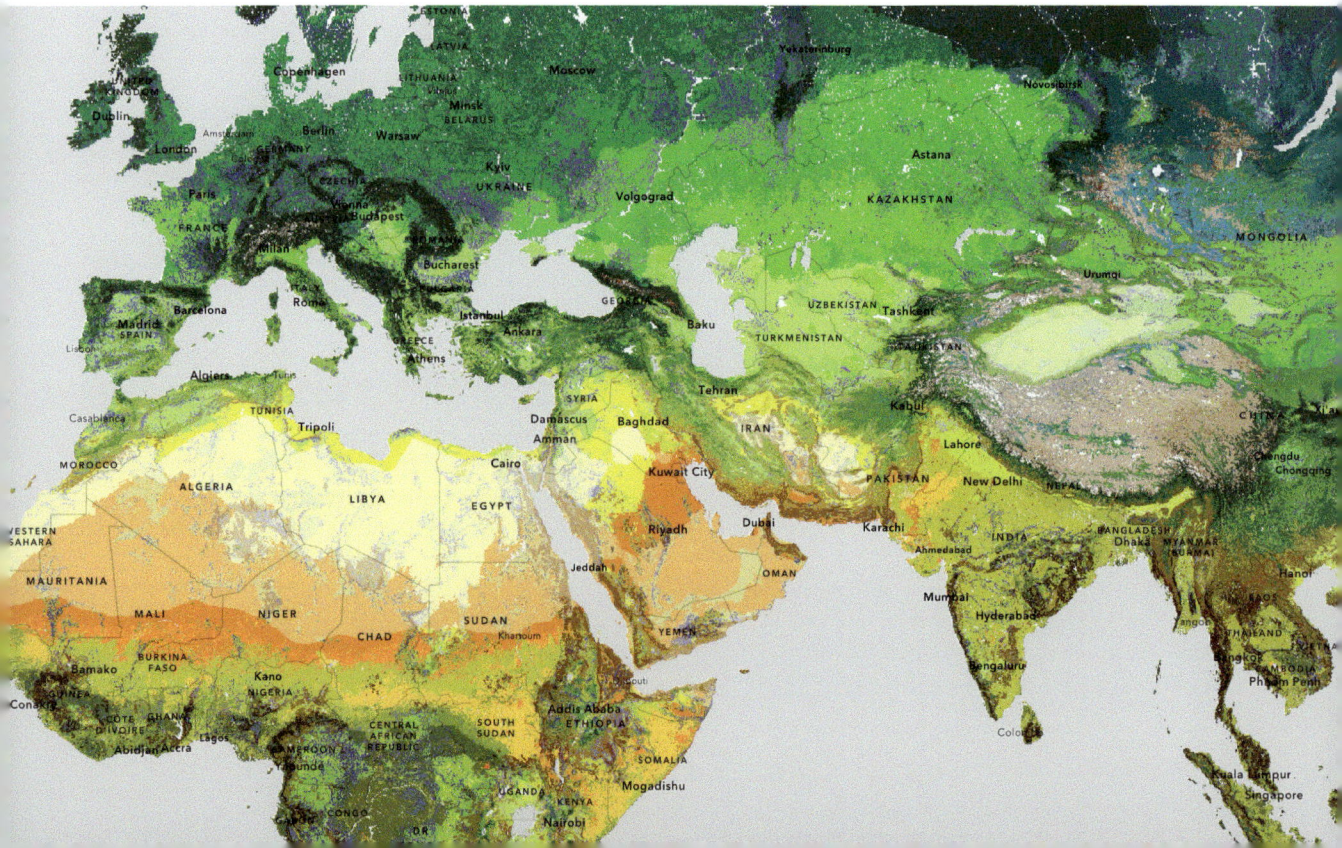

many thousands of geospatial professionals. I was thrilled to witness the growth of a pretty good idea into a world-class product, thanks to the talent and dedication of dozens of colleagues. At both organizations I rubbed elbows with extraordinarily talented and committed people—explorers, photographers, writers, cartographers, illustrators, and developers.

My work over these rewarding years can be boiled down to two verbs. At National Geographic, the word was *inform*. At Esri, it's *enable*. Throughout my stints in Nat Geo's art and map divisions, I created maps and graphics that informed millions of readers about El Niño, Earth history, the human immune system, the Big Bang theory, biodiversity, the solar system, and countless other topics. It was fun to imagine young people poring over our maps and artwork, just as I had done as a child, leafing through my grandmother's collection of magazines going back to 1928.

At Esri, we frequently create stories, often in collaboration with nonprofits and NGOs, which use our storytelling medium to inform people about

El Niño diagram, *National Geographic Magazine*, February 1984.

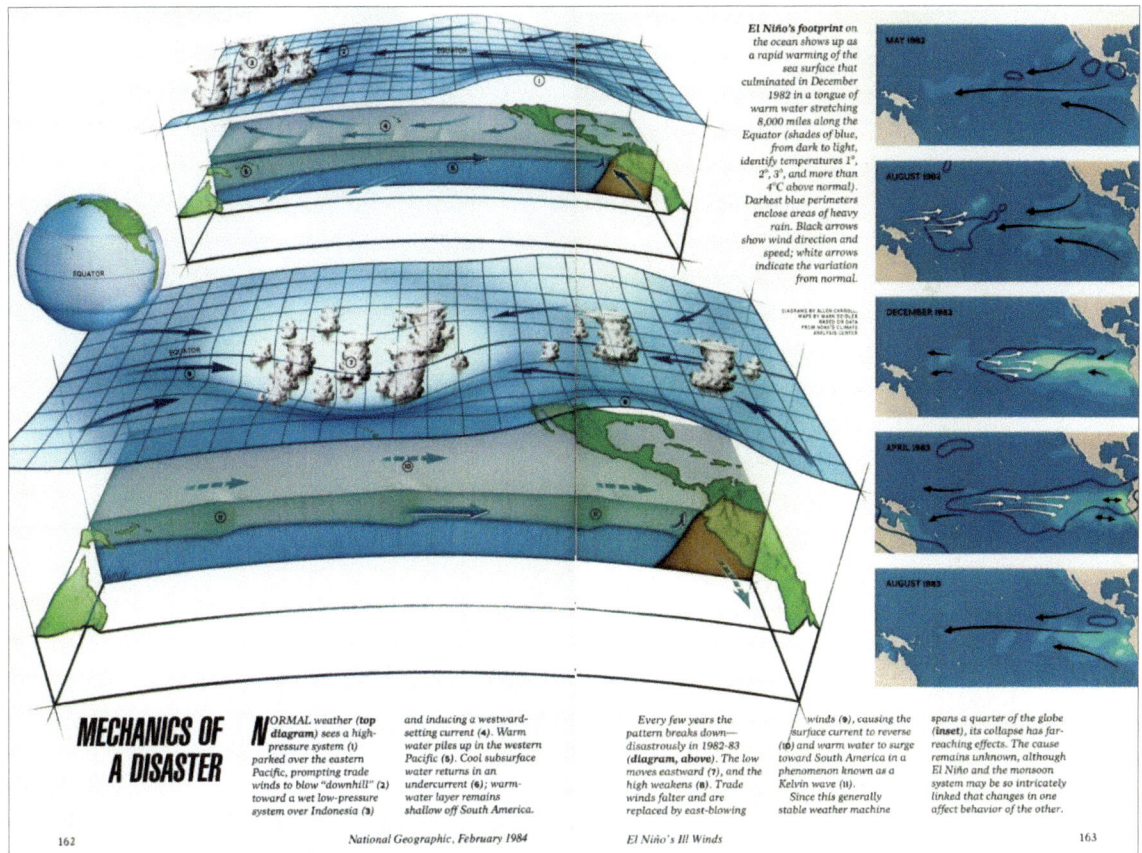

El Niño's footprint on the ocean shows up as a rapid warming of the sea surface that culminated in December 1982 in a tongue of warm water stretching 8,000 miles along the Equator (shades of blue, from dark to light, identify temperatures 1°, 2°, 3°, and more than 4°C above normal). Darkest blue perimeters enclose areas of heavy rain. Black arrows show wind direction and speed; white arrows indicate the variation from normal.

**MECHANICS OF A DISASTER**

NORMAL weather (top diagram) sees a high-pressure system (1) parked over the eastern Pacific, prompting trade winds to blow "downhill" (2) toward a wet low-pressure system over Indonesia (3) and inducing a westward-setting current (4). Warm water piles up in the western Pacific (5). Cool subsurface water returns in an undercurrent (6); warm-water layer remains shallow off South America.

Every few years the pattern breaks down—disastrously in 1982-83 (diagram, above). The low moves eastward (7), and the high weakens (8). Trade winds falter and are replaced by east-blowing winds (9), causing the surface current to reverse (10) and warm water to surge toward South America in a phenomenon known as a Kelvin wave (11).

Since this generally stable weather machine spans a quarter of the globe (inset), its collapse has far-reaching effects. The cause remains unknown, although El Niño and the monsoon system may be so intricately linked that changes in one affect behavior of the other.

important topics. But our much greater impact has been to enable—to make storytelling tools that help people create their own stories by the millions. Perhaps our greatest source of gratification comes from our frequent encounters with beautiful, inspiring, and informative stories created by individuals and organizations using our tools to articulate an important cause or envision a better future. That, in a larger sense, is what Esri is all about: providing GIS software and services that enable organizations to manage their activities and to understand the world around them. ArcGIS StoryMaps has enabled GIS professionals to more effectively and vividly describe their work and share the insights that their GIS analyses have revealed.

## Stories in the classroom

A second source of gratification comes from the degree to which ArcGIS StoryMaps has been used by educators and students. I'm a bit ashamed to admit this, but during our early work developing these products, we gave little thought to their utility as educational tools. We were soon thrilled to discover that educators were using them for instructional purposes. Much more thrilling has been the widespread creation of ArcGIS StoryMaps narratives by students, often as alternatives to traditional research papers. The intuitive nature of the ArcGIS StoryMaps builder experience—especially for young "digital natives"—has resulted in the creation of hundreds of thousands of student-produced stories. Although we don't have the numbers to prove it, we're confident that many of those students have discovered geography and GIS through their early use of ArcGIS StoryMaps and have gone on to further geographic pursuits. As the creation and use of place-based narratives continues to grow, it's exciting to think that we're helping expose a generation of young minds to the kind of holistic thinking that is an integral part of the geographic approach, and, more generally, that we're helping open those minds to the richness, the wonder, and the complexity of the world.

## The future of ArcGIS StoryMaps

This book, alas, is already out-of-date. Place-based storytelling in general and ArcGIS StoryMaps in particular continue to evolve despite publishing deadlines. There's no doubt that, by the time you're reading these pages, we will have added new storytelling capabilities to the ArcGIS StoryMaps list of features. We'll continue to make incremental enhancements; meanwhile, I can't resist speculating about what new forms of place-based storytelling may occur in the longer term.

One aspect of story tours that has always been challenging, at least until now, is that these tours aren't as useful as they could be when you're *in* the place. I'd love to see stories that use augmented reality (AR) to label and describe features, reveal historical photos, and place models of future structures in landscapes. I'd love to see automatic alerts as I approach landmarks or cross boundaries. I'd love to see a virtual dotted line appear on the ground in front of me, leading me to the next location. I'd love to have a bird's-eye, or drone's-eye, view follow me around so I can see myself on the map and see what's beyond a line of trees or cluster of buildings. But there are pros and cons to be weighed: Will I need to be staring into my smartphone or wearing a bulky headset?

As an extension of those AR stories, I'd like to find out what place-based stories might be near my current location. Imagine a city government or chamber of commerce creating walking tours for every neighborhood of a city. You query "nearby stories," get a list, pick one, and receive walking directions to starting points.

## Artificial intelligence (AI)

Artificial intelligence (AI) is on everyone's mind these days; there's no doubt that it will play an important part in future enhancements to ArcGIS StoryMaps. As this book went to press, designers and developers on our team were refining concepts and prototypes for AI-driven functions. Among the prototypes is a "writing assistant" that would augment the existing rich-text editor to check for spelling and grammar mistakes; other features might include rephrasing sentences and summarizing wordy passages. A separate function will assist in making stories more accessible by generating alternative text for machine screen readers. Additional goodies will likely include translation, readability assessments, estimated reading times, and more. Further in the future: mapping and image assistants.

It's important to note that AI-related functions will be opt-in; that is, account administrators will need to explicitly choose to include—or not include—AI-powered items. Educators, for instance, might opt out in order to require students to do all their own writing and editing.

## The story of the Earth

Every day, thousands of people are creating stories with ArcGIS StoryMaps. Individuals are aggregating dozens of stories into collections, hubs, galleries, and websites. Organizations are creating styles and themes to manage hundreds of stories. Authors are creating templates that allow others to create kindred stories, both visually and editorially. Universities are encouraging students to create stories by the tens of thousands.

The notion may be pure fantasy, or it might be megalomaniacal, but could this growing world of place-enabled stories ultimately self-organize into an übernarrative of our planet?

GIS users with a broad perspective of their work have long imagined that geographic information systems—and systems of systems—might evolve

The Earth is cloaked in an intricate fabric of ecosystems.

into a kind of nervous system for the planet. Not a *central* nervous system, but an emerging, distributed system of perception, analysis, knowledge, and understanding that could help us toward—and that could be essential to—our finding a way to live sustainably and equitably on this little planet of ours.

Earlier in this book, we discussed how storytelling is a basic part of being human and that we learn about and understand the world, in no small degree, through creating, sharing, and consuming stories. GIS gives us the tools to gather information, manage infrastructure, understand the present, and predict the future. Maybe place-based stories play a complementary role: informing and inspiring us, tapping into our innate navigational skills, using our collective spatial cognition, providing the emotional stimulus to move us to action—action that is guided by the insights we gain through GIS.

We happen to live at the precise moment in human history—and in the 4.5 billion-year history of the earth—when our activities threaten to irreparably harm the natural systems on which we depend for our survival. Yet, at this same moment, we're seeing an increase in willful ignorance, disinformation, and xenophobia. Our collective vision, or lack thereof, will determine our own fate. We have perhaps a generation or two to muster the resolve to save ourselves and heal a reeling planet. One can hope that, with a little help from tools that chart the quickening heartbeat of a threatened Earth and stories that inspire us to action as planetary stewards, we can map a path to a healthy and harmonious future.

# Acknowledgments

## Esri

**ArcGIS StoryMaps team members current and former**

David Asbury

Lee Bock

Jon Bowen

Cristina Cañizares

Gustavo Cardenas

Heidi Daulton

Warren Davison

Ross Donihue

Ashley Du

Rupert Essinger (in memoriam)

William Hackney

Greyson Harris

Mark Harrower

Sharon Kitchens

Gregory L'Azou

Andria Olson

Amelia Semprebon

Stephen Sylvia

Cooper Thomas

Michelle Thomas

Liz Todd

Hannah Wilber

Lara Winegar

...and Sathya Prasad, Owen Evans, and the entire ArcGIS StoryMaps dev team

**Assistance with this book**

Christian Harder

Mark Henry

Sharon Kitchens

David Oberman

Victoria Roberts

Alycia Tornetta

**Esri leadership**

Sean Breyer

Clint Brown

Jack and Laura Dangermond

Bern Szukalski

## National Geographic

Kevin Allen

Frank Biasi

Gil Grosvenor

Howard Paine (in memoriam)

## Other

Marjorie Hunt

Tess Carroll

Grace Carroll

Richard H. Goodwin (in memoriam)

# References

## Introduction
Carroll, Allen. "Deep Dupont." *Washington Post Magazine,* November 1, 1981.

Carson, Rachel. *Silent Spring.* Boston: Houghton Mifflin, 1962.

## Chapter 1
Dangermond, Jack. *The Power of Where: A Geographic Approach to the World's Greatest Challenges.* Redlands: Esri Press, 2024.

Valdés, Juan José. "About National Geographic Maps." *National Geographic.* www.nationalgeographic.com/maps/article/about-maps

Vonnegut, Kurt. "Kurt Vonnegut on the Shapes of Stories." YouTube, uploaded by David Comberg, October 30, 2010. www.youtube.com/watch?v=oP3c1h8v2ZQ

## Chapter 2
Richardson, Jim. "Photography, Conservation, Communications," presented at WildSpeak, September 2019.

Tobler, W. R. "A Computer Movie Simulating Urban Growth in the Detroit Region." *Economic Geography* 46 (1970): 234–40. https://doi.org/10.2307/143141

Wurman, Richard Saul. *Information Anxiety.* New York: Doubleday, 1989.

## Chapter 3
Baumann, Jim. "Geospatial Brain Power." *ArcWatch,* October 2019. www.esri.com/about/newsroom/arcwatch/geospatial-brain-power

Green, Adam. Interview by Allen Carroll, February 2024.

Harvard Health. "Right Brain/Left Brain, Right?" *Harvard Health Blog,* March 24, 2022. www.health.harvard.edu/blog/right-brainleft-brain-right-2017082512222

Kolvoord, Bob. Interview by Allen Carroll, January 2024.

Maguire, Eleanor A. et al. "London Taxi Drivers and Bus Drivers: A Structural MRI and Neuropsychological Analysis." *Hippocampus* vol. 16, no. 12 (2006): 1091–101. doi:10.1002/hipo.20233

Moser, May-Britt and Edvard I. Moser. "The Brain's GPS Tells You Where You Are and Where You've Come From." *Scientific American,* January 1, 2016. www.scientificamerican.com/article/the-brain-s-gps-tells-you-where-you-are-and-where-you-ve-come-from

Zak, Paul J. "Why Your Brain Loves Good Storytelling." *Harvard Business Review*, October 28, 2014. https://hbr.org/2014/10/why-your-brain-loves-good-storytelling

## Chapter 4

Allen, Kevin. Interview by Allen Carroll, February 2024.

Anders, William. "AS08-14-2383." NASA Image and Video Library, December 24, 1968. https://images.nasa.gov/search?q=AS08-14-2383&page=1&media=image,video,audio&yearStart=1920&yearEnd=2024

Brewer, Cynthia A. *Designing Better Maps: A Guide for GIS Users*. 3rd ed. Redlands: Esri Press, 2024.

Jabr, Farris. "The Reading Brain in the Digital Age: Why Paper Still Beats Screens." *Scientific American*, November 1, 2013. www.scientificamerican.com/article/the-reading-brain-in-the-digital-age-why-paper-still-beats-screens

National Geographic. *Atlas of the World*. 7th ed. National Geographic Society, 1999.

*National Geographic Magazine*. "The Two Koreas." Vol. 204, no. 1, July 2003.

## Chapter 5

National Geographic Society. *Historical Atlas of the United States*. National Geographic Society, 1988.

## Chapter 6

Buckley, Aileen, Allen Carroll, and Clint Brown. "The Mapmaker's Mantra." *ArcGIS Blog*, February 4, 2022. www.esri.com/arcgis-blog/products/arcgis-online/mapping/mapmakers-mantra

Dangermond, Jack, Jim Fallows, and Gary Knell. "UC Central Live: Maps & Storytelling." Panel discussion at the 2021 Esri User Conference. Video. Vimeo. https://vimeo.com/579465644

Monmonier, Mark. *How to Lie with Maps*. 2nd ed. Chicago: University of Chicago Press, 1996.

## Chapter 7

Harrower, Mark. "How to Get the Most from Interactive Maps." ArcGIS Blog, June 05, 2020. www.esri.com/arcgis-blog/products/arcgis-storymaps/mapping/how-to-get-the-most-from-interactive-maps

## Chapter 8

Carroll, Allen, and Ross Donihue. "These 11 Verbs Will Make You a More Effective Storyteller and Help You Excel at a Medium That's a Basic Part of Being Human." The Esri ArcGIS StoryMaps team, March 10, 2022. https://storymaps.arcgis.com/stories/00c17c5f8f8d41b5bbe0cdb 552117of4

Du, Ashley. "Effective Strategies for Engaging with Your Audience." *ArcUser*, Spring 2024. https://www.esri.com/about/newsroom/arcuser /effective-strategies-for-engaging-with-your-audience

Gather, Janine, and Wes Bernstein. "A GIS Showcase at Bucknell University for Esri." Mediaspace at Bucknell University. https://mediaspace. bucknell.edu/media/a+GIS+Showcase+at+Bucknell+University+for+ ESRI/1_m9xc1rvi

National Geographic Magazine. "Mechanics of a Disaster." Vol. 165, no. 2, February 1984.

Wilbur, Hannah. "Planning and Outlining Your Story Map: How to Set Yourself Up for Success." *ArcGIS Blog,* June 16, 2019. www.esri.com /arcgis-blog/products/arcgis-storymaps/sharing-collaboration /planning-and-outlining-your-story-map-how-to-set-yourself-up-for- success


Portions of this book were inspired by previous writings, including:

ArcGIS StoryMaps Collection. *Maps, Minds, and Stories.* https://storymaps. arcgis.com/collections/5451e979461a455ea583c0dd418543cf

Carroll, Allen. "Memory, History, and Place: Exploring Personal Geographies." *ArcGIS Blog,* April 25, 2018. www.esri.com/arcgis-blog /products/story-maps/mapping/memory-history-and-place-exploring- personal-geographies

# Credits

## Introduction

**ix, top.** Photo courtesy of Allen Carroll.

**ix, bottom.** Images courtesy of Rand McNally and Company & Rand McNally Collection. (1960). *Shell Indianapolis: Featuring metropolitan and downtown Indianapolis Speedway area.* (1960 edition.) [Map]. Rand McNally & Co.

**x.** Photo courtesy of Allen Carroll.

**xi.** Cartography by Allen Carroll, provided courtesy of the Linda Lear Center for Special Collections and Archives, Connecticut College.

**xiii.** Cartography by Allen Carroll; "How to Look at Deep Dupont" by Allen Carroll was first published in *The Washington Post* on November 1st, 1981.

## Chapter 1

**xvi-1.** Map from "Little Hero on the Prairie" story by the Esri ArcGIS StoryMaps team; Cartography by Warren Davison; Data from iNaturalist, Esri, and RESOLVE Ecoregions and Biomes dataset.

**2.** Giovanni Domenico Tiepolo, *The Storyteller*, 1773, Public domain, via Wikimedia Commons.

**9.** Images from "Little Hero on the Prairie" story by the Esri ArcGIS StoryMaps team; Photos by © Farshid, © Roberto, and © 潔 丹野 / Adobe Stock; Cartography by Warren Davison; Data from Central Grasslands Roadmap Initiative, U.S. Fish & Wildlife Service, Bioscience, an Ecoregions-Based Approach to Protecting Half the Terrestrial Realm DOI: https://doi.org/10.1093/biosci/bix014.

**12.** Maps from "America's Mental Health Crisis, Mapped" story by Este Geraghty, Esri Chief Medical Officer, and the Esri ArcGIS StoryMaps team; Cartography by Cooper Thomas; Data from University of Wisconsin Population Health Institute and Robert Wood Johnson Foundation.

**13.** Images courtesy of "Satellites and Seeds" story by Alcis GIS; Photo courtesy of Seeds for Development; Map © Alcis 2023, basemap © Esri.

**14, top left.** Image from "Thirteen Ways of Looking at the Grand Canyon" story by the Esri ArcGIS StoryMaps team; Data from Billingsley, G.H., et al., U.S. Geological Survey (USGS).

**14, top right.** Image from "Thirteen Ways of Looking at the Grand Canyon" story by the Esri ArcGIS StoryMaps team; Data from USGS, TNC, Esri.

**14, bottom left.** Image from "Thirteen Ways of Looking at the Grand Canyon" story by the Esri ArcGIS StoryMaps team; Data from USGS, Esri.

**14, bottom right.** Image from "Thirteen Ways of Looking at the Grand Canyon" story by the Esri ArcGIS StoryMaps team; Data from USDA Forest Service Forest Inventory & Analysis Program; Science by Barry T. Wilson (USFS); Cartography by Emily Meriam.

**15.** Images courtesy of "Living Territories" story by Amazon Conservation Team.

**16.** Images from "Living in the Age of Humans" story by the Esri ArcGIS StoryMaps team; Cartography by David Asbury, Ross Donihue, and Cooper Thomas; Data from © ESA Climate Change Initiative - Land Cover led by UCLouvain (2017).

**17.** Images courtesy of "Recovering Lost Crab Pots of the Salish Sea" story by Jefferson MRC & Sea Dragons with support from NW Straits Initiative, WDFW & NOAA; Photo by Monica Montgomery (*top*).

**18.** Images from "Combining Crowdsourced Data and ArcGIS StoryMaps" story by the Esri ArcGIS StoryMaps team.

**19, top.** Image from "Endemic Species" story by the Esri ArcGIS StoryMaps team; Photos by © Kevin Sloniecki, © Eric Isselée, © Sync, © Jiri Prochazka, © Edyta, © Galyna Andrushko, © Cecílio, © slowmotiongli, © Stillfx, © fiermanmuch, © Victor Tyakht, © Daniel Lamborn, © Todd Winner, © bennytrapp / Adobe Stock.

**19, bottom.** Image from "Endemic Species" story by the Esri ArcGIS StoryMaps team; Photos by © Eric Isselée, © Stillfx, © fiermanmuch, © atosan, © fotomaster / Adobe Stock; Cartography by Warren Davison.

**20, top.** Image from "Residential Redevelopment in the West Don Lands" story by the Esri ArcGIS StoryMaps team; Photo by © Mark Lotterhand / Adobe Stock.

**20, bottom left.** Image from "Residential Redevelopment in the West Don Lands" story by the Esri ArcGIS StoryMaps team; Data from City of Toronto, contains information licensed under the Open Government Licence – Toronto.

**20, bottom right.** Image from "Residential Redevelopment in the West Don Lands" story by the Esri ArcGIS StoryMaps team; Photo by © Curioso.Photography / Adobe Stock.

**21, top right and center.** Images courtesy of "The Burning of Greenwood" story by Brenna Maloney; Photo by J.E. Purdy & Co., Boston, 1899, courtesy of Library of Congress (*top right*), *The Tulsa Star*. (Tulsa, OK), Dec. 11 1920, courtesy of Library of Congress (*center*), © Greg Kelton / Adobe Stock (*bottom*).

**22, 1.** Image from "Storytelling for a Sustainable World" story collection by the Esri ArcGIS StoryMaps team; Cartography by John Nelson; Imagery from "NASA Blue Marble" GEBCO: "GEBCO Compilation Group (2021)" and Natural Earth: "Natural Earth".

**22, 2.** Photo by Nabil Naidu / Unsplash.

**22, 3.** Photo by Pixabay / Pexels.

**22, 4.** Photo by Getty Images / Unsplash.

**22, 5.** Photo by RF._.studio / Pexels.

**22, 6.** Photo by Pixabay / Pexels.

**22, 7.** Photo by Red Zeppelin / Pexels.

**22, 8.** Photo by Josh Olalde / Unsplash.

**22, 9.** Photo by ThisisEngineering / Unsplash.

**22, 10.** Photo by Julie Ricard / Unsplash.

**23.** Images courtesy of "Objects of Wonder" story by Smithsonian National Museum of Natural History; Mogul Emerald Necklace, National Gem Collection, Smithsonian Institution. Gift of Mrs. Madeleine H. Murdock. Photo by Ken Larsen and digitally enhanced by SquareMoose (*top*) and James Di Loreto, Smithsonian (*bottom*).

**24.** Images courtesy of "Geospatial Conservation at The Nature Conservancy" story by The Nature Conservancy; Cartography by Chris Bruce/TNC; Photos by Vienna Saccomanno/TNC (*left*) and Jonathan MacKay/TNC (*right*).

**25.** Images from "Share your EarthPlaces" story by the Esri ArcGIS StoryMaps team.

**26.** Images courtesy of "Scraping the Heavens" story by Cooper Thomas; Photos courtesy of Cooper Thomas; Data from Natural Earth, Esri.

**28-29.** Images courtesy of "Living Territories" story by Amazon Conservation Team; Data from Agencia Nacional de Mineria, ANNA platform.

**30.** Images courtesy of "1,001 Novels: A Library of America" story by Susan Straight; Photo by Felisha Carrasco.

**31.** Images courtesy of "1,001 Novels: A Library of America" story by Susan Straight; Photo by Douglas McCulloh.

## Chapter 2

**32-33.** Map from "American Agriculture by the Numbers" story by the Esri ArcGIS StoryMaps team; Cartography by Cooper Thomas; Data from Esri, US Census Bureau, US Department of Agriculture - National Agriculture Statistics Service.

**35.** Image from "Welcome to Palm Springs" story by the Esri ArcGIS StoryMaps team; Photo courtesy of Mark Harrower.

**36, top.** Image from "The Lands We Share" story by the Esri ArcGIS StoryMaps team; Data from USGS, Esri.

**36, bottom.** Image from "An Introduction to Sea Ice" story by the Esri ArcGIS StoryMaps team; Cartography by Ross Donihue; Data from National Snow and Ice Data Center, University of Colorado, Boulder.

**39.** Images courtesy of "University Libraries as Providers of GIS Services: A Guide" story collection by David Cowen.

**40, top and bottom.** Photo courtesy of Allen Carroll.

**40, center.** Photo by © AboutLife / Adobe Stock.

**41.** Images courtesy of "Choking on Convenience" story by iLCP. Text by Meg Severide, Visuals & Storytelling Manager, The International League of Conservation Photographers (iLCP); Photos by © Sergio Izquierdo (*top left and bottom right*), © Randall Rosales (*top right*), and © Emanuele Biggi (*bottom left*), The International League of Conservation Photographers (iLCP).

**43.** Images from "American Agriculture by the Numbers" story by the Esri ArcGIS StoryMaps team; Cartography by Cooper Thomas; Data from Esri, US Census Bureau, US Department of Agriculture - National Agriculture Statistics Service.

**45.** Images courtesy of "A River Interrupted" story by Charles River Watershed Association; Photo by Sean MacNamara; Illustration designed by Julia Hopkins.

**46, top.** Image courtesy of "World's Longest Mule Deer Migration: Red Desert to Hoback" story; Photo by Gregory Nickerson, Wyoming Migration Initiative, University of Wyoming.

**46, bottom.** Image courtesy of "World's Longest Mule Deer Migration: Red Desert to Hoback" story; Map courtesy of *Wild Migrations: Atlas of Wyoming's*

*Ungulates*. Oregon State University Press © 2018 University of Wyoming and University of Oregon.

**47, top.** Image courtesy of "World's Longest Mule Deer Migration: Red Desert to Hoback" story; Map courtesy of *Wild Migrations: Atlas of Wyoming's Ungulates*. Oregon State University Press © 2018 University of Wyoming and University of Oregon.

**47, bottom left.** Images courtesy of "World's Longest Mule Deer Migration: Red Desert to Hoback" story; Photos by Brian Remlinger (*bottom left*) and Leon Schatz and Gregory Nickerson, Wyoming Migration Initiative, University of Wyoming (*bottom right*).

**47-48.** Images courtesy of "Malaria on the Frontlines" story by United Nations Foundation's United to Beat Malaria campaign.

## Chapter 3

**50-51.** Map courtesy of "From Naptown to Indy" story by Allen Carroll; Imagery basemap by the ArcGIS Living Atlas team.

**52.** Cartography by Allen Carroll.

**54, left.** Image by Henry Vandyke Carter, Public domain, via Wikimedia Commons.

**55.** Images courtesy of Green Lab, Georgetown University.

**56.** Image courtesy of Green Lab, Georgetown University.

**58.** Images from "Doing Conservation on the Ground" story by the Esri ArcGIS StoryMaps team; Photos by Sharon Kitchens (*left*) and Frank Cangelosi (*right*).

**59.** Images courtesy of "Aerial Odysseys: Bird Migration in the Americas" story by National Audubon Society; Story written by Melanie Smith/Audubon and the Esri ArcGIS StoryMaps team; Photo by Dagny Gromer; Cartography by Warren Davison; Swainson's Hawk map migration tracks from Salt Lake City Airport Wildlife Mitigation Team (2020; Movebank study 317134988), and seasonal ranges derived from eBird at the Cornell Lab of Ornithology and BirdLife International; Data from the Bird Migration Explorer/Audubon (Smith et al. 2022).

**60.** Images from "The Diversity of Life" story by the Esri ArcGIS StoryMaps team; Cartography by Cooper Thomas; Data from BioScience, An Ecoregions-Based Approach to Protecting Half the Terrestrial Realm DOI: https://doi.org/10.1093/biosci/bix014; Photo by Zdenek Machacek / Unsplash.

**62.** Images courtesy of "The Lines That Shape Our Cities" story by Digital Scholarship Lab at the University of Richmond, The Science Museum of Virginia, and Esri; Photo Adobe Stock.

**63.** Images courtesy of "The Lines That Shape Our Cities" story by Digital Scholarship Lab at the University of Richmond, The Science Museum of Virginia, and Esri; Photo courtesy of California Digital Library; Data from USGS national elevation dataset; Historical map from Mapping Inequality Project.

**64.** Images courtesy of "The Old Man of the Mountain in 3D" story by Matthew Maclay; 3D model by Matthew Maclay.

**65.** Images courtesy of "The Old Man of the Mountain in 3D" story by Matthew Maclay; Photo by The Old Man of the Mountain Legacy Fund's Richard F Hamilton Collection; 3D model by Matthew Maclay.

## Chapter 4

**66-67.** Map courtesy of "The Two Koreas" story by the Esri ArcGIS StoryMaps team; Cartography by Cooper Thomas; Data from European Commission, Joint Research centre and Google Earth users Panda and Planeman.

**68.** Image courtesy of National Geographic Society.

**69.** Image courtesy of National Geographic Society.

**70.** Images courtesy of National Geographic Society.

**71.** Images courtesy of National Geographic Society.

**72.** Image courtesy of National Geographic Society.

**73.** Image courtesy of National Geographic Society.

**75.** Images courtesy of "Peaks and Valleys" story by the Esri ArcGIS StoryMaps team.

**77.** Maps courtesy of "The River Roads of India" story by Paul Salopek; Cartography by Cooper Thomas; Data from Natural Earth, World Wildlife Fund, World Resources Institute.

**79.** Images courtesy of National Geographic Society.

**80.** Images from "The Two Koreas" story by the Esri ArcGIS StoryMaps team; Photos by Joseph A. Ferris III and © SeanPavonePhoto / Adobe Stock; Cartography by Cooper Thomas; Data from United States Military Academy.

**81.** Images from "The Two Koreas" story by the Esri ArcGIS StoryMaps team; Cartography by Cooper Thomas; Data from United States Military Academy.

**83.** Image courtesy of National Geographic Society.

**85.** Photo courtesy of NASA.

**86.** Images from "Battles of the American Civil War" story by the Esri ArcGIS StoryMaps team; Photos by Kurz and Allison, Public domain, via Wikimedia Commons (*top*) and Alexander Gardner, Public domain, via Wikimedia Commons (*bottom*).

**87.** Images from "Battles of the American Civil War" story by the Esri ArcGIS StoryMaps team; Image courtesy of Library of Congress.

**88-89.** Images from "Travels with Godzilla" story collection by Allen Carroll; Photos by Allen Carroll.

## Chapter 5

**90-91.** Map from "The Human Reach" story by the Esri ArcGIS StoryMaps team; Cartography by David Asbury and Cooper Thomas; Data from NASA's Earth Observatory.

**92.** Image courtesy of National Geographic Society.

**93.** Images courtesy of National Geographic Society.

**95.** Images courtesy of National Geographic Society.

**96.** Image courtesy of National Geographic Society.

**97.** Images courtesy of National Geographic Society.

**99.** Images courtesy of National Geographic Society.

**100.** Image courtesy of National Geographic Society.

**101.** Image courtesy of National Geographic Society.

**102.** Image courtesy of Allen Carroll.

**103.** Images from "Beating the Odds: A Year in the Life of a Piping Plover" story by National Audubon Society Enterprise GIS; Photos by Walker Golder; © 2025 National Audubon Society, Inc. All rights reserved.

**104, top and center.** Images from "Geography, Class, and Fate: Passengers on the Titanic" story by the Esri ArcGIS StoryMaps team.

**104, bottom left and right.** Images from "Twister Dashboard: Exploring Four Decades of Violent Storms" story by the Esri ArcGIS StoryMaps team; Data from NOAA severe report database.

**105.** Image from "A Walk on the High Line" story by Allen Carroll; Photos by Allen Carroll.

**106, top.** Image from "Spyglass on the Past: Washington DC 1851 and Today" story by the Esri ArcGIS StoryMaps team; Historical map from Library of Congress, Geography and Map Division.

**106, center.** Image from "Spyglass on the Past: New York City 1836 and Today" story by the Esri ArcGIS StoryMaps team; Historical map from David Rumsey Map Collection, David Rumsey Map Center, Stanford Libraries.

**106, bottom.** Image from "Top 10 Most Visited National Parks in 2012" story by Esri; Photos by National Park Service.

**108, top.** Image from "Katrina +10: A Decade of Change in New Orleans" story by the Esri ArcGIS StoryMaps team; Data from 2006 American Community Survey 1-year Estimates, State-to-State Migration Flows, NHC, NOAA, NWS.

**108, center.** Image from "The Ten Most Damaging Hurricanes in the History of the United States" story by the Esri ArcGIS StoryMaps team; Data and photo from NOAA.

**108, bottom.** Image from "Mapping Renewable Energy around the World" story by the Esri ArcGIS StoryMaps team.

**109, top.** Image from "The 2014 Esri User Conference" story by the Esri ArcGIS StoryMaps team; Photos by Allen Carroll.

**109, bottom.** Image from "San Diego Shortlist" story by the Esri ArcGIS StoryMaps team; Photos by Rupert Essinger, © f11photo / Adobe Stock, and © MelissaMN / Adobe Stock.

**109, bottom right.** Image courtesy of Rupert Essinger.

**110, top.** Image from "Uprooted" story by the Esri ArcGIS StoryMaps team; Photo by © UNHCR/Viktor Pesenti; Copyright © UNHCR. All rights reserved.

**110, center.** Map from "Uprooted" story by the Esri ArcGIS StoryMaps team; Data from UNHCR Refugee Population Statistics Database.

**110, bottom.** Image from "Uprooted" story by the Esri ArcGIS StoryMaps team; Photos by © UNHCR/Christopher Herwig (*left*) and © UNHCR/Mark Henley (*right*); Copyright © UNHCR. All rights reserved.

**112.** Images from "The Surprising State of Africa's Giraffes" story by the Esri ArcGIS StoryMaps team; Photo by Pawel Czerwinski / Unsplash.

**113, top.** Images from "Our Favorite Esri Stories of 2019" story collection; Photos by © Acronym / Adobe Stock, Gabby Salazar, Cooper Thomas, Clint Patterson / Unsplash, Pawel Czerwinski / Unsplash, and Indianapolis Public Library digital collection.

**113, bottom right.** Images from "Living in the Age of Humans" story collection by the Esri ArcGIS StoryMaps team; Photos by Jezael Melgoza / Unsplash, Dave Hoefler / Unsplash, Jplenio / Pixabay, Zdenek Machacek / Unsplash.

**114.** Images from "ArcGIS StoryMaps: An Introduction" story by Ashley Du; Photo by © Piotr Krzeslak / Adobe Stock.

**115.** Images from "U.S. Wildfire Snapshot" story by the Esri ArcGIS StoryMaps team; Photo by the National Interagency Fire Center.

**116.** Photos courtesy of Allen Carroll.

**118–119.** Images courtesy of the Ralph Rinzler Folklife Archives and Collections, Center for Folklife and Cultural Heritage, Smithsonian Institution; Photos by Tom Pich.

**120–121.** Images courtesy of City of Irvine.

## Chapter 6

**122–123.** Map courtesy of "Coastal Flooding" story by NOAA/NOS, Old Dominion University, Esri; Data from NOAA CO-OPS, Old Dominion University.

**125.** Images courtesy of "At Nature's Crossroads" story by The Nature Conservancy.

**126, bottom.** Image from "World Heritage in Danger" story by the Esri ArcGIS StoryMaps team; Photo by Gilles Mairet, CC BY-SA 3.0.

**127.** Image from "Coastal Flooding" story by NOAA/NOS, Old Dominion University, Esri; Photos by NOAA (*left*) and Department of Foreign Affairs and Trade, CC BY 2.0 (*right*); Data from NOAA CO-OPS, Old Dominion University.

**129.** Images from "Out of Eden Walk: Milestones Map" story by Paul Salopek; Photos by Paul Salopek.

**130.** Image from "Abandoned Islands" story by the Esri ArcGIS StoryMaps team; Photos by © Dmitry / Adobe Stock (*top*) and Bertramz, CC BY 3.0 (*bottom*).

**132.** Images courtesy of "Living Territories" story by Amazon Conservation Team.

**133, top left.** Image from "Justice Deferred" story by the Esri ArcGIS StoryMaps team; Photo from U.S. Office of War Information.

**133, top right.** Image from "Justice Deferred" story by the Esri ArcGIS StoryMaps team; Photos by Ansel Adams, from the Library of Congress.

**133, center left.** Image from "Justice Deferred" story by the Esri ArcGIS StoryMaps team; Cartography by Cooper Thomas and Clare Trainor; Data from Ben Pease/Japantown Atlas; Photo by Clem Albers, courtesy of National Archives Catalog.

**133, center right.** Image from "Justice Deferred" story

by the Esri ArcGIS StoryMaps team; Photo by Dorothea Lange, courtesy of the Library of Congress.

**133, bottom left and right.** Images from "Justice Deferred" story by the Esri ArcGIS StoryMaps team; Cartography by Cooper Thomas and Clare Trainor; Data from War Relocation Authority/internmentarchives.org.

**134.** Map from "Mapping the Thanksgiving Harvest" story by the Esri ArcGIS StoryMaps team; Cartography by David Asbury; Data from USDA Census of Agriculture.

**135.** Image from "Age of Megacities" story by the Esri ArcGIS StoryMaps team; Data from United Nations, Department of Economic and Social Affairs, Population Division. Online Data.

**136, top and center.** Images from "Age of Megacities" story by the Esri ArcGIS StoryMaps team; Data from Lincoln Institute of Land Policy.

**136, bottom left and right.** Images from "Age of Megacities" story by the Esri ArcGIS StoryMaps team; Data from Esri, WorldPop.

**137, top right.** Image from "Wealth Divides" story by the Esri ArcGIS StoryMaps team; Image from Esri Imagery basemap.

**137, center right and bottom.** Image from "Wealth Divides" story by the Esri ArcGIS StoryMaps team; Data from U.S. Census Bureau.

**140.** Images courtesy of "Dante's *Inferno*" story by Josef Münzberger.

**141.** Map courtesy of "Dante's *Inferno*" story by Josef Münzberger; Cartography by Josef Münzberger.

**142.** Image courtesy of "Okavango Explore" story by EarthViews & Blue Water GIS; Photos by Jake Sorenson (*top*) and Kostadin Luchansky, National Geographic Society (*center*).

**143.** Images courtesy of "Okavango Explore" story by EarthViews & Blue Water GIS; Cartography by Blue Water GIS.

## Chapter 7

**144–145.** Map from "Walking China's Antique Roads" by Paul Salopek; Cartography by Cooper Thomas.

**146.** Image from "Planning and Outlining Your Story" Esri blog post by Hannah Wilber; Photo by Hannah Wilber.

**147, bottom center.** Image courtesy of "On Foot in the

Path of the Silk Road" story by Paul Salopek; Photo by Paul Salopek.

**147, bottom top right.** Image from "Hot Numbers" story by the Esri ArcGIS StoryMaps team.

**147, bottom right.** Image courtesy of "Something's Buzzing!" story by Maisong Francis, City of Brooklyn Park, MN; Photo by Adege / Pixabay.

**148.** Image from "What's That Bug" story by the Esri ArcGIS StoryMaps Team; Illustration by Warren Davison.

**149.** Image courtesy of "When Rains Fell in Winter" story by Philip Burgess & Irina Wang; Photo by Andrei Golovnev.

**150.** Images from "Charging Across the Country" story by the Esri ArcGIS StoryMaps team; Cartography by Ashley Du and David Asbury; Data from U.S. Department of Energy's Alternative Fuels Data Center.

**151.** Image courtesy of "Bear Necessities" story; Story by Archbold Biological Station; Photo by Carlton Ward, Jr.; Cartography by Angeline Meeks, Archbold Biological Station.

**152, top left and right.** Images courtesy of "Walking China's Antique Roads" by Paul Salopek; Cartography by Cooper Thomas.

**152, center left and right.** Images courtesy of "Walking China's Antique Roads" by Paul Salopek; Photos by Paul Salopek.

**152, bottom left.** Image courtesy of "Walking China's Antique Roads" by Paul Salopek; Cartography by Cooper Thomas; Data from China Historical GIS v6 (2016).

**152, bottom right.** Image courtesy of "Walking China's Antique Roads" by Paul Salopek; Cartography by Cooper Thomas; Data from Peter Bol.

**153.** Images from "(Farm) Animal Planet" story by the Esri ArcGIS StoryMaps team; Chart data from Food and Agriculture Organization of the United Nations. FAOSTAT. Crops and livestock products. Latest update: 2017. Accessed: 2019. Map data from Gilbert, M., Nicolas, G., Cinardi, G. et al. Global distribution data for cattle, buffaloes, horses, sheep, goats, pigs, chickens and ducks in 2010. Sci Data 5, 180227 (2018) doi:10.1038/sdata.2018.227.

**155, left.** Images courtesy of "Of All the Fish in the Sea" story by Aaron Koelker; Photo (*center*) courtesy of NOAA National Marine Fisheries Service; Data from Maryland Department of Natural Resources.

**155, right.** Images courtesy of "The Diverse Prague" story by the Institute of Planning and Development (IPR Prague).

**156, top.** Image courtesy of "What You CAN'T See in the Tennessee River" story by Kellie Crye Ward.

**156, bottom.** Image courtesy of Allen Carroll.

**157.** Images courtesy of "The Half-Earth Project" story by Half-Earth Project, Map of Life, Yale Center for Biodiversity and Global Change, E.O. Wilson Biodiversity Foundation.

**159.** Images from "Resurfacing the Past" by the Esri ArcGIS StoryMaps team; Cartography by Cooper Thomas and Ross Donihue; Data from Paul Heersink.

**160.** Image from "Climate Migrants" story collection by the Esri ArcGIS StoryMaps team; Photos by © Jiri, © ilfotokunst, © logoboom, © Vladimir Melnikov / Adobe Stock, and Mil.ru, CC BY 4.0, via Wikimedia Commons.

**161.** Images courtesy of "From Dying to Wild Again" story by Tompkins Conservation; Photo by Matias Rebak, courtesy of Rewilding Argentina.

**162-163.** Images courtesy of "Divots and Dashboards" story by Warren Davison and the Esri ArcGIS StoryMaps team; Illustrations by Warren Davison.

**164.** Images from "Hot Numbers" story by the Esri ArcGIS StoryMaps team; Photo by Daria Lisovtsova / Unsplash (*center left*), Pawan Sharma / Unsplash (*bottom right*).

**165.** Images from "Hot Numbers" story by the Esri ArcGIS StoryMaps team; Photo by Christian Baluja / Unsplash; Cartography by Cooper Thomas; Data from NSIDC/NASA.

## Chapter 8

**166-167.** Map from "Spyglass on the Past: New York City 1836 and Today" story by the Esri ArcGIS StoryMaps team; Image from David Rumsey Map Collection, David Rumsey Map Center, Stanford Libraries.

**171.** Image from ArcGIS StoryMaps gallery; Photos by David Medeiros, Stanford Geospatial Center and Andria Olson, © Curioso.Photography / Adobe Stock, IUCN, and NIFC.

**172.** Images courtesy of Hannah Wilber.

**173.** Image courtesy of "Homecoming" story by Dr. Rae Wynn-Grant; Photos by OrnaW / Pixabay (*top*), Carolyn

Barnwell (*center*), and © Uryadnikov Sergey / Adobe Stock (*bottom*).

**175.** Image from "Getting Started with ArcGIS StoryMaps" story by Allen Carroll and the Esri ArcGIS StoryMaps team.

**177, top right.** Image courtesy of Allen Carroll.

**177, bottom.** Images courtesy of "Mushroom Finder" story by Ross Donihue; Photos by © Susie Hedberg, © nkarol, © Ivan, © James, © fotografiecor, and © Hye young / Adobe Stock.

**179, right.** Image from "The Surprising State of Africa's Giraffes" story by the Esri ArcGIS StoryMaps team; Photo by Pawel Czerwinski / Unsplash.

**180.** Image from "Thirteen Ways of Looking at the Grand Canyon" story by the Esri ArcGIS StoryMaps team; Data from © ESA WorldCover project 2020 / Contains modified Copernicus Sentinel data (2020) processed by ESA WorldCover consortium.

**181.** Image from "Urban Africa" story by the Esri ArcGIS StoryMaps team; Photo by Jorge Tung / Unsplash (*left*) and Sergey Pesterev / Unsplash (*right*).

**182.** Images from "What's That Bug" story by the Esri ArcGIS StoryMaps Team; Illustrations and cartography by Warren Davison; Data from New York State Integrated Pest Management Program.

**184, left.** Images from "The Diversity of Life" story by the Esri ArcGIS StoryMaps team; Cartography by Cooper Thomas; Data from BioScience, An Ecoregions-Based Approach to Protecting Half the Terrestrial Realm DOI: https://doi.org/10.1093/biosci/bix014.

**184, top center and right.** Images from "The Diversity of Life" story by the Esri ArcGIS StoryMaps team; Photos by Александр Лещёнок, CC BY-SA 4.0, via Wikimedia Commons and Elizabeth Morgan / Unsplash.

**184, bottom center and right.** Images from "The Diversity of Life" story by the Esri ArcGIS StoryMaps team; Photos by © salparadis and © Lukas / Adobe Stock.

**185, top.** Images from "André versus Capability" story by Allen Carroll and the Esri ArcGIS StoryMaps team; Photos by Allen Carroll (*left*) and Blenheim Palace, CC BY-SA 4.0, via Wikimedia Commons (*right*).

**185, right center.** Image from "André versus Capability" story by Allen Carroll and the Esri ArcGIS StoryMaps team; "Portrait of André Le Notre" by Carlo Maratta and "Capability Brown" by Nathaniel Dance-Holland.

**185, bottom.** Image from "Age of Megacities" story by the Esri ArcGIS StoryMaps team; Data from Lincoln Institute of Land Policy.

**186.** Image from "Answering the Call" story by the Esri ArcGIS StoryMaps team; Cartography by Cooper Thomas.

**187-188.** Images from "Six Scottish Hikes, Six Tour Formats" story by Allen Carroll and the Esri ArcGIS StoryMaps team; Photos by Allen Carroll.

**190, top left.** Image from "Grace and Delight" story by Will Hackney and the Esri ArcGIS StoryMaps team; Photo by William Henry Jackson, Library of Congress (*left*) and Will Hackney (*right*).

**190, top center.** Image from "Grace and Delight" story by Will Hackney and the Esri ArcGIS StoryMaps team; Photo courtesy of David Rumsey Map Collection, David Rumsey Map Center and Library of Congress, Geography and Map Division.

**190, bottom left.** Image from "Grace and Delight" story by Will Hackney and the Esri ArcGIS StoryMaps team; Photo courtesy of David Rumsey Map Collection, David Rumsey Map Center.

**191, left.** Image from "Bombing Missions of the Vietnam War" story by Cooper Thomas and the Esri ArcGIS StoryMaps team; Photos by Capt. Robert H. Glaves, U.S. Navy, Public domain, via Wikimedia Commons, Mark Wilkins, Public domain, via Wikimedia Commons, and Kimlong Meng, CC BY-SA 4.0, via Wikimedia Commons; Data from Department of Defense's THOR program.

**191, right.** Image courtesy of "Green Oranges & Land" story by Aaron Koelker.

**192.** Image from "Share Your Earth Places" story by the Esri ArcGIS StoryMaps team.

**193.** Image from "Bombing Missions of the Vietnam War" story by Cooper Thomas and the Esri ArcGIS StoryMaps team; Photo by U.S. Navy, Public domain, via Wikimedia Commons; Data from Department of Defense's THOR program.

**194.** Image from "Age of Megacities" story by the Esri ArcGIS StoryMaps team; Map from Esri's world imagery basemap; Data from Esri, Maxar, Earthstar Geographics, and the GIS User Community.

**195.** Photo courtesy of Allen Carroll.

**196.** Image from "Thirteen Ways of Looking at the Grand Canyon" story by the Esri ArcGIS StoryMaps team; Data from © ESA WorldCover project 2020 /

Contains modified Copernicus Sentinel data (2020) processed by ESA WorldCover consortium.

**202.** Images courtesy of "The Ebony Project" story collection by The Ebony Project.

**203.** Images courtesy of "The Ebony Project" story collection by The Ebony Project; Photo by Vincent Deblauwe (*bottom right*).

**204-205.** Images from "Global Fishing Watch" story by Craig McCabe; Cartography by Emily Meriam; Data from Global Fishing Watch; Photos by © whitcomberd and © Gunnar E Nilsen / Adobe Stock.

## Epilogue

**206.** Images from "Living in the Age of Humans" story by the Esri ArcGIS StoryMaps team; Cartography by David Asbury, Ross Donihue, and Cooper Thomas; Data from © ESA Climate Change Initiative - Land Cover led by UCLouvain (2017).

**207.** Cartography by Allen Carroll; Image courtesy of National Geographic Society.

**210.** Images from "Living in the Age of Humans" story by the Esri ArcGIS StoryMaps team; Cartography by David Asbury, Ross Donihue, and Cooper Thomas; Data from © ESA Climate Change Initiative - Land Cover led by UCLouvain (2017).

# About Esri Press

Esri Press is an American book publisher and part of Esri, the global leader in geographic information system (GIS) software, location intelligence, and mapping. Since 1969, Esri has supported customers with geographic science and geospatial analytics, what we call The Science of Where. We take a geographic approach to problem-solving, brought to life by modern GIS technology, and are committed to using science and technology to build a sustainable world.

At Esri Press, our mission is to inform, inspire, and teach professionals, students, educators, and the public about GIS by developing print and digital publications. Our goal is to increase the adoption of ArcGIS and to support the vision and brand of Esri. We strive to be the leader in publishing great GIS books, and we are dedicated to improving the work and lives of our global community of users, authors, and colleagues.

**Acquisitions**

Stacy Krieg

Claudia Naber

Alycia Tornetta

Jenefer Shute

**Product Engineering**

Craig Carpenter

Maryam Mafuri

**Editorial**

Carolyn Schatz

Mark Henry

David Oberman

**Production**

Monica McGregor

Victoria Roberts

**Sales & Marketing**

Eric Kettunen

Sasha Gallardo

Beth Bauler

**Contributors**

Christian Harder

Matt Artz

**Business**

Catherine Ortiz

Jon Carter

Jason Childs

# Related titles

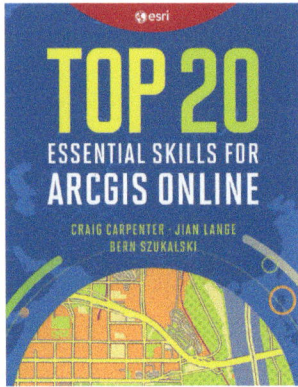

**Top 20 Essential Skills for ArcGIS Online**

Craig Carpenter, Jian Lange, and Bern Szukalski

9781589487802

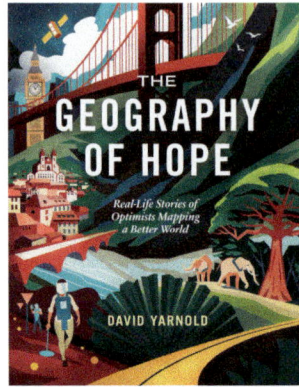

**The Geography of Hope**

David Yarnold

9781589487413

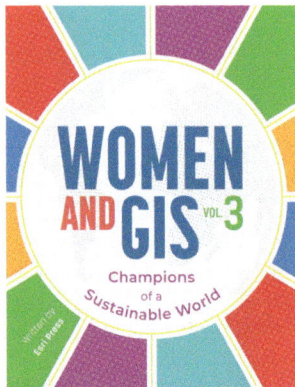

**Women and GIS, volume 3**

Esri Press

9781589486379

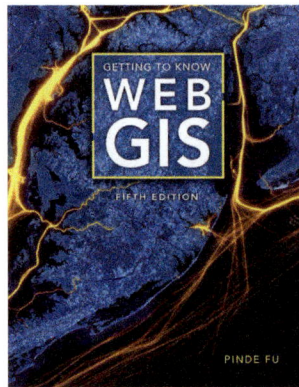

**Getting to Know Web GIS, fifth edition**

Pinde Fu

9781589487277

For more information about Esri Press books and
resources, or to sign up for our newsletter, visit

**esripress.com**.